Images of Modern America

SATURN V ROCKET

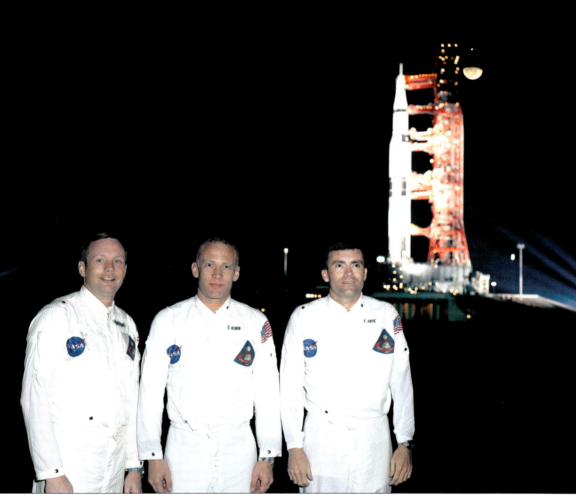

This unusual crew photograph shows the Apollo 8 backup crew of, from left to right, Neil Armstrong, Buzz Aldrin, and Fred Haise on December 10, 1968, in front of the SA-503 Apollo 8 Saturn V and the target, the moon. (KSC-68C-8008. Kipp Teague.)

FRONT COVER: The launch of Apollo 14 on January 31, 1971. (AS14-71PC-152. J.L. Pickering.)

UPPER BACK COVER: The launch of Apollo 15 on July 26, 1971. (AS15-71HC-1107. J.L. Pickering.);

LOWER BACK COVER: (from left to right) Apollo 11 first stage on the Pearl River barge on August 29, 1968 (68-69295. NASA.), Dr. Wernher von Braun (left) and Dr. Kurt Debus (right) at the roll out of the SA-500F Saturn V on May 25, 1966 (KSC-66PC-110. Kipp Teague.), F-1 engines being installed in the S-IC-T stage on March 30, 1965 (65-21876. NASA.)

Images of Modern America

SATURN V
ROCKET

ALAN LAWRIE
FOREWORD BY ED STEWART II, INTRODUCTION BY MIKE JETZER

ARCADIA
PUBLISHING

Published by Arcadia Publishing
Charleston, South Carolina

Printed in the United States of America

Library of Congress Control Number: 2016934548

For all general information, please contact Arcadia Publishing:
Telephone 843-853-2070
Fax 843-853-0044
E-mail sales@arcadiapublishing.com
For customer service and orders:
Toll-Free 1-888-313-2665

Visit us on the Internet at www.arcadiapublishing.com

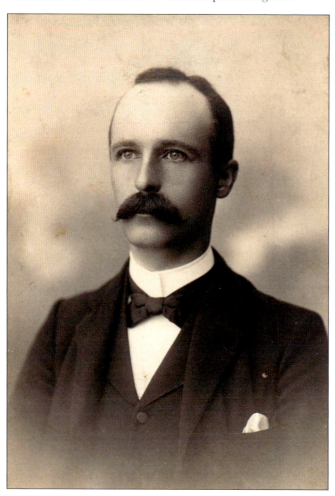

This book is dedicated to my great-grandfather Cecil Robert Hillman, who was a talented British mechanical engineer. He was born on January 18, 1868, at 28 Strada Rinella, Cottonera, Malta, and died on March 31, 1943, at Rua Maestro Chiafarelli 386, Sao Paulo, Brazil. He immigrated to Brazil on March 20, 1896, and became chief mechanical engineer of the Sao Paulo Railway. In his spare time, he produced the first planisphere of the southern hemisphere and wrote books on astronomy and general relativity. (Author's collection.)

CONTENTS

ACKNOWLEDGMENTS

After the caption for each photograph I have added the official photograph number if one exists. I have also added the source(s) for each photograph. I gratefully acknowledge the support of the following people who contributed to this book.

Arlene Royer of the National Archive and Records Administration in Atlanta scanned and provided photographs from the original negatives held at NARA.

Mike Jetzer (heroicrelics.org) from Wisconsin wrote the introduction to this book and proofread all of my captions. He also provided various video capture images. In addition, Mike is now the host of the collection of Douglas Aircraft Company photographs, previously located on Phil Broad's cloudster website, which originally came from the Santa Monica Museum of Flight.

Ed Stewart II, director of exhibits and curation at the US Space & Rocket Center in Huntsville, wrote the foreword.

J.L. Pickering, president of Retro Space Images, provided most of the photographs in chapters 5 and 6. Kipp Teague from the Project Apollo Archive provided a number of images, as did Jim Porter, Dick Serrano, Terri Pennello, Vince Wheelock, Doug Galloway, Dieter Zube, Tom Usciak, and Andy Clark.

David Concannon and Geoff Nunn were involved with the recovery and display of the recovered F-1 engine parts.

Special thanks to Dr. David Baker, whose Apollo and Saturn articles in the 1970s *Spaceflight* magazines inspired this teenage space enthusiast.

Jeff Ruetsche and Liz Gurley provided wonderful support for this book at Arcadia Publishing.

To Olwyn Lawrie, all my love.

All author proceeds from the sale of this book will go to the US Space & Rocket Center in Huntsville.

—Alan Lawrie
Hitchin, England
March 2016

FOREWORD

The Saturn V is, in the mind of the layman, associated with Cape Canaveral. The mighty Apollo launch vehicle was truly a product of several states, with Florida and the cape being only the most public portion of its lifecycle. Rarely do people realize that California, Florida, Missouri, Alabama, and more were involved.

The Saturn V is, at heart, a rocket from the southeastern United States. Developed largely at Marshall Space Flight Center (MSFC) in Huntsville, Alabama, it is part of a legacy of technology that is unexpected in the Deep South. The design of the rocket was overseen by Dr. Wernher von Braun, who was part of the Operation Paperclip scientist group and the developer of the V-2 missile. Dr. von Braun refined the concept throughout his time working for the Army Ballistic Missile Agency (ABMA) at Redstone Arsenal. When he was appointed as the director of the newly formed National Aeronautics and Space Administration (NASA) Marshall Space Flight Center in 1960, work on the rocket began in earnest. Evolving the Saturn designs from the Redstone and Jupiter technologies, von Braun took President Kennedy's lunar mandate and built the most powerful launch vehicle ever flown.

The Army and ABMA helped edge what used to be known as the "Watercress Capital of the World" into new technology. But it wasn't until the Saturn V began to take shape that the city saw what would become a permanent change. Huntsville saw contractors from all over the country converge in areas surrounding Marshall and the arsenal. The Saturn V brought companies like Boeing, IBM, Lockheed, Northrop, and more. Cummings Research Park was formed to accommodate these companies and the new local ventures that formed to support the program. Now Cummings Research Park houses companies serving defense, aerospace, software development, and even the biotechnology industries. It is second in size only to Research Triangle Park in North Carolina.

The Saturn and its challenges also helped establish the University of Alabama in Huntsville. It was founded to support the education and training needs of the growing local aerospace industry. It provided opportunities for local students to learn these fields and then secure jobs at home. Renowned for its programs in engineering, it has grown to include atmospheric science, materials science, and more.

Above all, the Saturn V changed the culture of Huntsville. Along with von Braun's German team, people from all over the United States and with diverse cultural backgrounds came to work on the rocket. Most importantly, many stayed after the Apollo program ended. These people helped establish the Huntsville Museum of Art, the Huntsville Symphony Orchestra, the US Space & Rocket Center (USSRC), and Space Camp.

Huntsville built the rocket that put humans on the moon and a mark in history. But the Saturn V had a profound effect on its hometown. The museums, the wildly varied food scene, and the diverse population are the cultural offspring of the Saturn V and truly make Huntsville, in every sense of the phrase, "The Rocket City."

—Ed Stewart II
Director of Exhibits & Curation
US Space & Rocket Center
Huntsville, Alabama

INTRODUCTION

I write this from my home in a suburb of Milwaukee, Wisconsin, a city that played a vital role in producing and launching the Saturn V.

Some people may know that Boeing produced the S-IC (first) stage at the Michoud Assembly Facility (MAF) in New Orleans; North American Aviation (NAA) manufactured S-II (second) stages in Seal Beach, California; Douglas Aircraft Corporation (DAC) fabricated the S-IVB (third) stage in Huntington Beach, California; Rocketdyne produced the F-1 and J-2 rocket engines in Canoga Park, California; and that Huntsville, Alabama, was the home of both the Marshall Space Flight Center (MSFC) (which oversaw the entire rocket-building enterprise) and the IBM facility that manufactured the Instrument Unit (IU).

Others may know that Rocketdyne performed J-2 acceptance testing and very early F-1 development testing at the Santa Susana Field Laboratory (SSFL) near Simi Valley, California, but that F-1 testing was soon moved to the less-populous Air Force Rocket Propulsion Laboratory (AFRPL), located between Lancaster and Rosamond in California's Mojave Desert. They may also be familiar with DAC's Sacramento Test Operations (SACTO), where S-IVB stages were test-fired, or the Mississippi Test Facility (MTF), at which the S-IC and S-II stages were acceptance tested.

But few people are likely to cite Milwaukee as having a role in producing the Saturn V. Yet without Milwaukee's contributions, the Saturn V would literally have never made it to the launch pad.

The Milwaukee suburb of South Milwaukee was home of the Bucyrus-Erie Company, a manufacturer of large mining machines. A NASA official's knowledge of this equipment led to Bucyrus landing a contract to design the crawler-transporter that would carry the Saturn V from the Kennedy Space Center's (KSC) Vehicle Assembly Building (VAB) to the launchpad. Although Bucyrus designed the crawler, NASA's competitive bidding process meant that the construction contract went to the Marion Power Shovel Company of Marion, Ohio.

Ladish Co., a metal forgings and fittings manufacturer located in Cudahy (another Milwaukee suburb), produced DC-6 steel, prized for its high strength and high temperature characteristics. The Thiokol Chemical Corporation's production facility in Elkton, Maryland, chose Ladish DC-6 steel from which to manufacture the case of the S-IC retro-rocket. The S-IC had eight of these retros, which slowed the spent stage after it burned out, thereby ensuring a safe distance between it and the rest of the launch vehicle before the S-II stage ignited to continue the journey to orbit.

Milwaukee was also home to the A.C. Spark Plug Division of General Motors Corporation, which helped produce the Apollo Command Module guidance system, but since this book is about the Saturn V proper, I'll not dwell on spacecraft systems.

But Milwaukee is far from unique in its role in helping to manufacture and launch the Saturn V. It has been estimated that, at its peak, more than 20,000 industrial firms employing more than 350,000 people were producing equipment for the Saturn/Apollo program. In 1962–1963, as construction and modification of facilities to support the Saturn V program was ramping up, a report noted that NASA awarded over $1 billion in contracts to firms in 46 states, with 67 percent of the contracts going to small businesses. While this includes all NASA contracts, not just Saturn V work, it demonstrates how cities all over America were contributing to the nation's space efforts.

Boeing manufactured the S-IC stage at NASA's MAF in New Orleans. Michoud originally built World War II cargo vessels on an assembly line. After the war, it was modified to manufacture plywood cargo planes, and in 1951, Chrysler leased it to produce engines for Sherman and Patton tanks. Chrysler later used the facility to build first stages for the smaller Saturn I and IB rockets.

Although Michoud was a huge facility, with over 1.8 million square feet, Boeing could not simply move in and immediately begin building S-IC stages. Mason-Rust (a joint venture of Rust Engineering Company of Pittsburgh and the Mason & Hanger-Silas Mason Company of Lexington, Kentucky), which provided support services such as security, medical and fire protection, photographic services, plant maintenance and repair, food service, custodial, and utilities, began a program of updating the facility to accommodate manufacture of the oversized rocket stage.

Mason-Rust hired the Gurtler-Hebert Company of New Orleans to remove the false steel framework in the ceiling of the manufacturing area. Welding and Manufacturing Company, also of New Orleans, modified trusses in the plant to increase their weight-bearing capability to facilitate the movement of components by overhead crane. The Haddad Construction Corporation, of Slidell, Louisiana, built a 40-foot-tall, 1,100-foot-long wall to separate the Saturn I/IB and Saturn V manufacturing areas.

Since final assembly of the stage would be performed vertically, it was necessary to construct a building for vertical assembly, hydrostatic test, and cleaning. The Vector Corporation of New Orleans provided design services and construction plans for the 214-foot-tall structure. Ross Corporation, also of New Orleans, was awarded the construction contract.

During Michoud's first three years of operation (including Chrysler's tenure building the earlier Saturn I/IB stages), $41 million was spent as dozens of companies, from the Gulf Coast and beyond, modified existing structures and constructed new facilities. After Hurricane Betsy damaged buildings in September 1965, Tri-State Roofing Company of Knoxville and J.A. Jones Construction Company of Charlotte made the necessary roofing, structural, electrical, and glass repairs.

The S-IC stage consisted primarily of various aluminum alloys and tempers. Alcoa, the Aluminum Company of America, provided 400,000 pounds of aluminum from plants in Alcoa, Tennessee; Davenport, Iowa; Cleveland, Ohio; Lafayette, Indiana; and Massena, New York. This was whittled down to 171,047 pounds as it was transformed from its raw form to stage structures such as propellant tanks, an inter-tank, a forward skirt, and thrust rings, held together by aluminum weld wire and rivets.

Gore segments, pie-shaped pieces that form the curved tank ends, were manufactured by Boeing's Wichita, Kansas, plant. After initial cutting and forming, the segments underwent a chemical milling process, developed by the Chemical Contour Corporation of Gardena, California, to remove excess material (and therefore excess weight) and achieve their final required thickness. Shipped to Michoud, the gore segments were joined to propellant tank walls using processes, tools, and enormous automated welding machines provided by Sciaky Bros. Inc. of Chicago.

The S-IC's liquid oxygen (LOX) tank was located forward of the fuel tank. Five seamless 40-foot-long, 25-inch-diameter tunnels, manufactured by Parsons Corporation of Traverse City, Michigan, via their Par-Form process, were installed to carry LOX through the fuel tank to the engines below.

Martin Company's Baltimore Division fabricated 17.5-foot-long, high-pressure, seamless aluminum helium bottles for installation inside the LOX tank. The helium would be used to pressurize the fuel tank during flight to replace the volume of fuel consumed by the engines. Sterer Engineering & Manufacturing Company of Los Angeles supplied check valves for the helium and other stage pneumatic systems.

To support the entire rocket on the launch pad and keep it restrained until full thrust was developed, the thrust structure incorporated four hold-down posts. The Wyman Gordon Company of Worcester, Massachusetts, supplied the 14-foot, 1,800-pound raw aluminum forgings (among the largest closed-die forgings ever fabricated at its plant) to Boeing's Wichita Branch, which then precision-machined each forging to its final shape before delivery to Michoud.

The S-IC stage, like the rest of the rocket, was heavily instrumented, allowing computers in the IU and on the ground to monitor the stage's performance. Electro Development Corporation of Lynnwood, Washington, supplied instrument amplifiers; measuring power supplies for instrumentation signals; data collection units; measurement systems and signal conditioners for pressure, force, strain, temperature, and vibration; and longitudinal accelerometers. Bendix's Instruments & Life Support Division of Davenport, Iowa, supplied propellant loading level sensors as well as sensors to detect liquid levels (providing a cutoff signal for the engines). Systron-Donner, of Concord, California, provided components for the propellant loading computer as well as servo accelerometers and pressure transducers. Power for the instrumentation, by these and dozens of other vendors, was supplied by silver-zinc batteries from the Electronics Division of Eagle-Picher Industries of Joplin, Missouri. Ground-based RCA 110A computers, supplied by RCA's Electromagnetic and Aviation Systems Division of Van Nuys, California, checked out the stages at Michoud and again after the stages were installed on the Mobile Launcher at KSC.

If the instrumentation showed a problem and it would be necessary to terminate the flight, the S-IC had two command receivers, supplied by Motorola's Government Electronics Division of Scottsdale, Arizona, to receive the self-destruct signal. The Link Ordnance Division of General Precision Inc. of Sunnyvale, California, supplied the Confined Detonating Fuse (CDF), which would activate linear shaped charges to cut the propellant tank walls, harmlessly releasing the stage's propellants, thus preventing a large explosion. Link Ordnance also provided the CDF pyrogen initiators, which ignited the S-IC's retro-rockets (the ones fabricated from Milwaukee's Ladish steel) at the end of a normal S-IC flight.

Near the end of the manufacturing process, Boeing mounted five Rocketdyne F-1 engines to the S-IC. In flight, the engines were gimballed by servo actuators manufactured by Moog Inc. of East Aurora, New York. Johns-Manville, of New York City, provided Micro-Quartz, a high-temperature fibrous insulation, to insulate the gimbal actuator arms from the engines' nearly 6,000 degree Fahrenheit exhaust gases.

Upon completion of manufacture, the stage was rolled out of Michoud on a transporter (the Allied Engineering and Development Company of Alameda, California, manufactured the wheel units; the American Machine and Foundry Company of Stamford, Connecticut, supplied the steering control and electrical power systems; and Harvey Aluminum Company of Terrence, California, provided steering actuators) to the Saturn Marine Dock (built by Farrell Construction Company Inc. of Memphis). There it was wheeled onto a retired Navy ocean-going barge, *Orion*, which had been modified so that its cargo space was large enough to accommodate the rocket stage by Diamond Manufacturing Company of Savannah, Georgia. The A.L. Mechling Barge Lines of Joliet, Illinois, towed, crewed, and maintained *Orion* and several other Saturn-related barges.

This parade of engineering and construction firms, vendors, and operators was repeated numerous times throughout the Saturn V program, not only by the other stage vendors but also for facilities required to test and launch the vehicle.

S-IC and S-II stages were test fired at MTF, a facility built specifically for Saturn V at a cost of $270 million. During the five years from its conception to its completion, over 48 major design and construction contracts were let, with subcontracts involving 200 to 300 additional companies.

KSC was also a new facility; its scale dwarfed MTF. The VAB, the Launch Control System, the launch pads, the crawlerway over which the Milwaukee-designed crawler-transporter traveled, and the myriad of support buildings required architect-engineering firms to design, construction firms to build, and electrical and computer companies to outfit. These firms came not only from Florida and the Gulf Coast but also from New York City; Fort Washington, Pennsylvania; Lynwood, California; Roseland, New Jersey; Tucson, Arizona; and Kansas City.

The Saturn V was not built by NASA or four or five major vehicle contractors. America, both small towns and large cities, built the Saturn V. The *Saturn V News Reference* lists "major subcontractors" (several contractors listed above did not make this list) from 142 cities in 26 states.

In a recent television interview, Apollo 17 astronaut Gene Cernan said, "We didn't go to the moon alone. We had every worker in this country who put a nut and bolt or a piece of wire in that spacecraft"—or launch vehicle.

—Mike Jetzer
heroicrelics.org

One

SATURN V FIRST STAGE

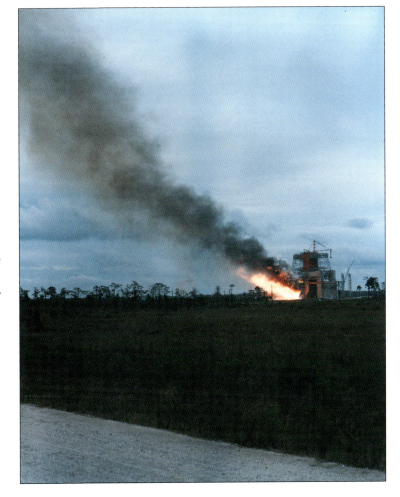

All of the Saturn V first stages from Apollo 9 (S-IC-4) onwards were test fired at the MTF, now known as NASA's Stennis Space Center. This photograph shows the 125-second test firing of the Apollo 10 (S-IC-5) stage on August 25, 1967. The test firings were used to check the functioning and performance of the complete first stage. (67-58559. Arlene Royer [NARA]/ Alan Lawrie.)

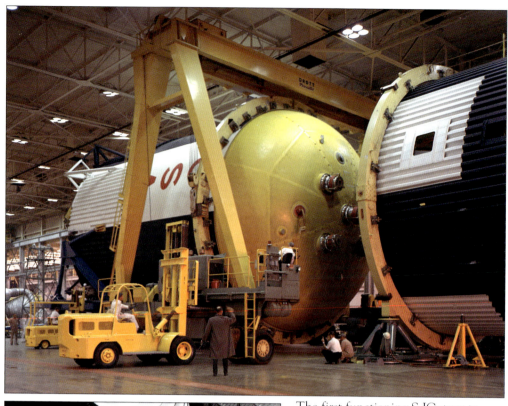

The first functioning S-IC stage to be built was designated S-IC-T. It was assembled at NASA's MSFC in Huntsville. This image shows the S-IC-T fuel tank (left) being prepared to be bolted to the inter-tank section in MSFC's building 4705 on December 18, 1964. In turn, the inter-tank would be bolted to the LOX tank. (64-16485. NASA.)

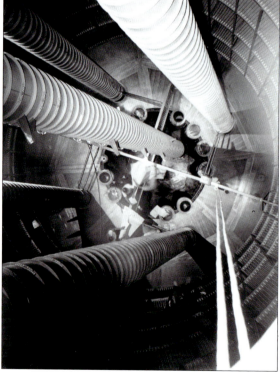

Five 42-foot tunnels fed liquid oxygen from the LOX tank down through the inside of the fuel tank to each of the five F-1 engines at the bottom of the first stage. This view inside the fuel tank shows workmen at the base of the tank, with the tunnels passing through the tank. (US Space & Rocket Center [USSRC].)

Following assembly of the S-IC-T stage at MSFC, the complete stage, minus F-1 rocket engines, was transported by road the short distance to MSFC's S-IC Static Test Stand in the West Test Area. This photograph, taken on March 1, 1965, shows the rear of the stage where the engines later would be attached. (65-20216. NASA.)

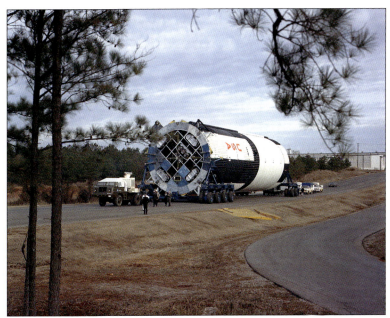

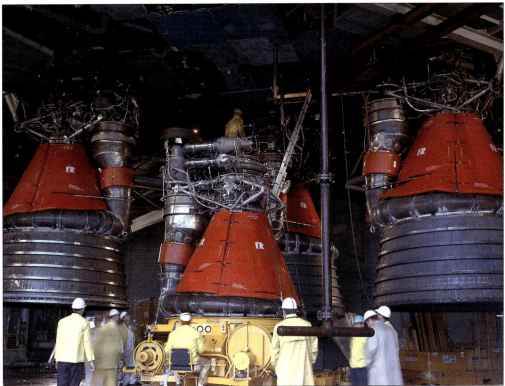

Once the S-IC-T stage had been secured vertically into the test stand, the five F-1 engines were attached at the base. In this March 30, 1965, photograph, three engines can be seen already attached, together with their nozzle extensions. A fourth engine, at this point without its nozzle extension, is attached to the vertical engine positioner, being lined up to be attached to the stage. (65-21876. NASA.)

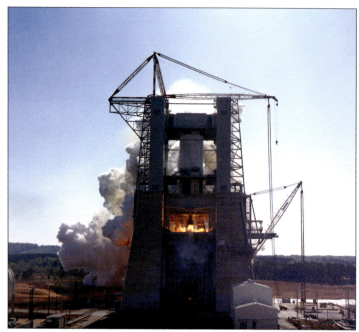

After three single-engine firings, the first firing of all five S-IC-T first-stage engines took place on April 16, 1965. The 6.5-second firing, generating a thrust of 7.5 million pounds, was a major milestone in the Saturn V program. The S-IC-T stage was test fired a total of 20 times. It can now be seen at the Kennedy Space Center in Florida. (65-22474. NASA.)

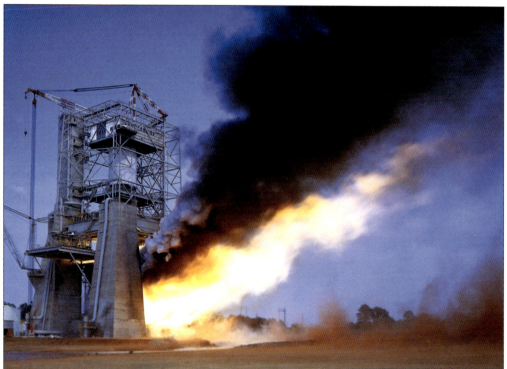

As well as being used for some of the test firings of the S-IC-T test stage, the purpose-built S-IC Static Test Stand was used for the acceptance test firings of the first three Saturn V flight vehicle first stages. Vibrations produced by the acoustics of the firings broke a number of windows in surrounding homes, and NASA had to deal with a number of insurance claims. This S-IC-T firing took place in May 1965. (65-23540. NASA.)

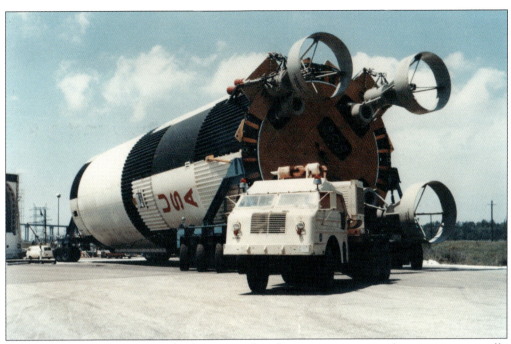

A Saturn V Dynamic Test Vehicle was produced comprising stages that were structurally representative but that need not be fired. The S-IC-D first stage had four mass simulators in place of engines (as seen here), plus a dummy engine. The S-IC-D stage was the first to be assembled by Boeing at MAF. The image shows the S-IC-D stage being loaded on the barge *Poseidon* at Michoud for the trip to MSFC on April 19, 1968. (68-66317. Arlene Royer [NARA]/Alan Lawrie.)

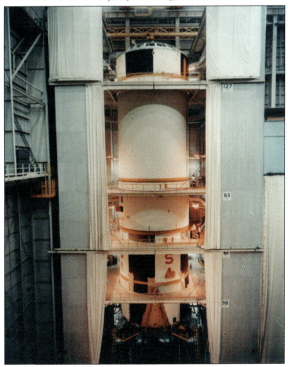

In addition to the S-IC-D test stage, all S-IC flight stages from S-IC-3 onwards were built by the industry contractor Boeing at Michoud. All other S-IC stages were built at MSFC by a combination of NASA and Boeing. This image shows the forward skirt being attached to the top of the S-IC-8 stage on August 15, 1967, completing the vertical assembly of the stage. (67-58554. Arlene Royer [NARA]/Alan Lawrie.)

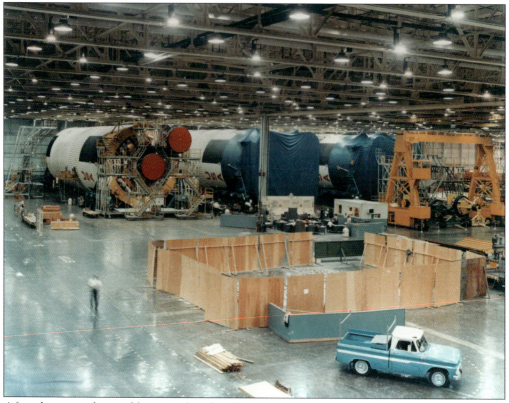

After the vertical assembly in Michoud's tall purpose-built tower, the S-IC stages were rotated horizontally for the integration of the five F-1 engines. Three S-IC stages are seen at various levels of assembly in this view in October 1967. From left to right, F-1 engines are being mated to the Apollo 13 (S-IC-8) stage, with Apollo 9 (S-IC-4) and Apollo 11 (S-IC-6) stages in storage. (67-60490. Arlene Royer [NARA]/Alan Lawrie.)

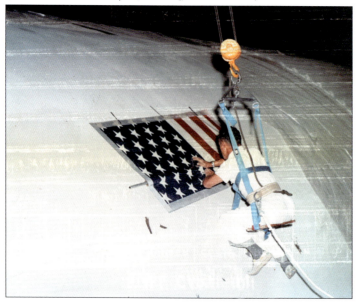

Each Saturn V had four large US flags attached around the circumference. Each flag was composed of silk-screen printed decal material in three sections. In this image, a technician is attaching a flag to the Apollo 10 S-IC-5 stage at Michoud on October 19, 1967. This previously unseen photograph is from a scan of the original negative, which has ink bleeding issues but is included because of its historical importance. (67-60475. Arlene Royer [NARA].)

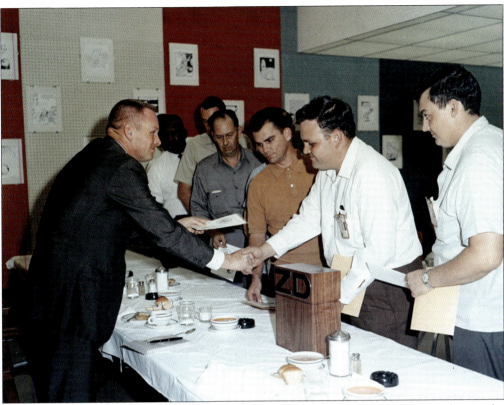

In order to boost morale, the astronauts who were to fly the Saturn V would frequently visit the sites of the major contractors and meet the workers. In April 1968, Neil Armstrong visited MAF and is seen autographing Roll of Honor folders for workers during a zero-defects awards banquet. (68-66320. Arlene Royer [NARA].)

During the manufacturing and testing activities at Michoud, each S-IC stage was regularly moved from one test cell to another. This photograph shows the Apollo 11 S-IC-6 stage being moved to Test Cell 2 of the Stage Test Building at Michoud in November 1967. In the background, Michoud's high bay, where the vertical assembly took place, can be seen. (67-61147. Arlene Royer [NARA]/Alan Lawrie.)

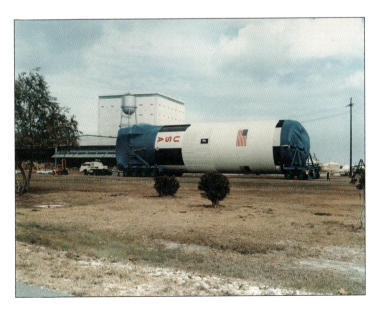

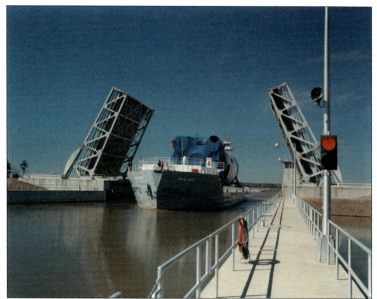

After each of the S-IC flight stages had been manufactured at Michoud, it was transported by open-top barge to and from the Mississippi Test Facility for a static firing test. The barges were pushed by tugboats. In this image, the Apollo 11 S-IC-6 stage is shown on board the barge *Pearl River* arriving at MTF on March 1, 1968. (68-64015. Arlene Royer [NARA]/Alan Lawrie.)

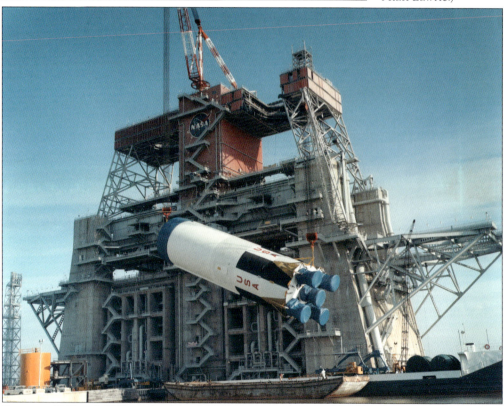

Between 1967 and 1970, each of the S-IC flight stages was test fired in the purpose-built B test stand at MTF. The test stand had two test positions, B-1 and B-2, each capable of supporting an S-IC test firing. In practice, the B-2 position was used in every case. This image shows the Apollo 16 S-IC-11 stage being installed in the test stand on May 13, 1969, prior to its first test firing. (69-75874. Arlene Royer [NARA]/Alan Lawrie.)

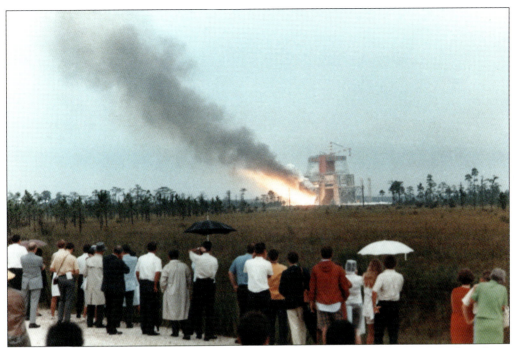

This rare and historic image shows the test firing of the Apollo 11 S-IC-6 first stage in the B-2 test stand at MTF on August 13, 1968. The successful 125-second firing of this stage paved the way for its use on the first moon landing mission. The original negative was located by the author and had not been published prior to this discovery. (68-68490. Arlene Royer [NARA]/Alan Lawrie.)

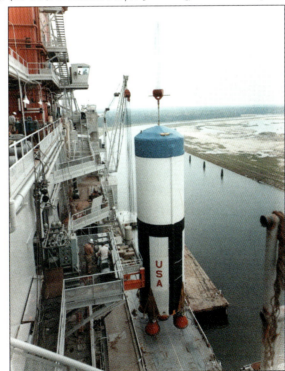

This photograph shows the Apollo 10 S-IC-5 stage being lowered out of the B-2 test stand onto a barge on September 11, 1967, after its test firing. After further checks at Michoud, each stage was shipped to KSC on one of two covered barges, *Orion* and *Poseidon*. (67-59485. Arlene Royer [NARA]/Alan Lawrie.)

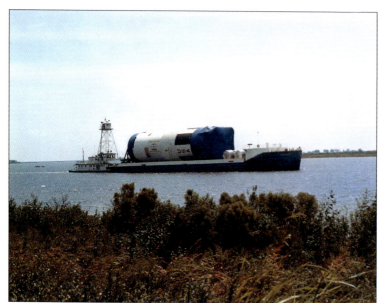

After its firing, the S-IC-6 stage was shipped back to MAF along the Pearl River on the barge *Pearl River*, as seen in this photograph taken on August 29, 1968. Apparently there were occasions when locals living on the riverside would take pot-shots at the passing valuable cargoes, with the flag as the target. (68-69295. NASA.)

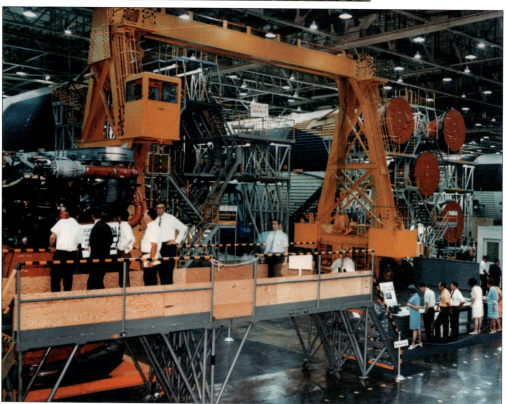

All the test firings at MTF were successful, with the exception of the Apollo 16 S-IC-11 firing on June 26, 1969, which was aborted after 97 seconds because of an engine fire. The failure happened because a small dust cover was not removed, which resulted in a leak in a hydraulic line. All the Michoud workers were required to examine the major consequence of a small quality control issue. (69-00732. Arlene Royer [NARA]/Alan Lawrie.)

Two

SATURN V SECOND STAGE

Each S-II second stage was test fired for around 370 seconds at MTF, whose A-1 and A-2 test stands were built for this purpose. The S-II used liquid hydrogen and liquid oxygen as propellants, and each stage had five J-2 engines with a thrust of 225,000 pounds, later upgraded to 230,000 pounds. This photograph, taken on November 27, 1967, shows the installation of the S-II-4 stage in the A-2 test stand. (67-61058. Arlene Royer [NARA]/ Alan Lawrie.)

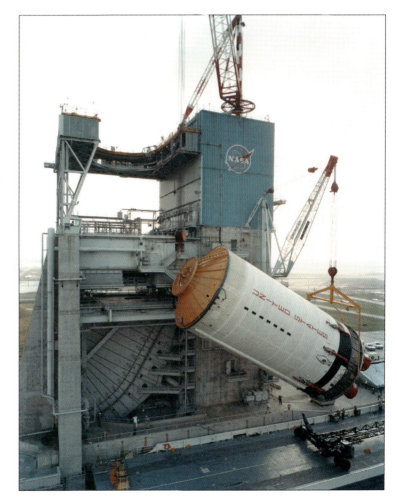

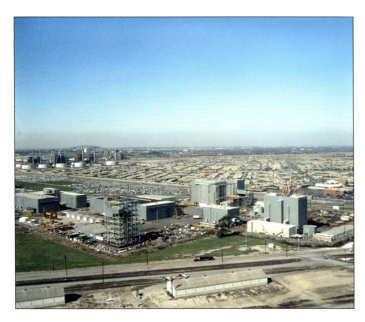

The S-II second stages were manufactured by North American Aviation, known as North American Rockwell (NAR) from 1967, at its plant in Seal Beach, California, which was leased from NASA. The Seal Beach facilities included a bulkhead fabrication building, 125-foot-high vertical assembly building, 116-foot-tall vertical checkout building, remote pneumatic test area, and a number of other structures. This view is from January 26, 1968. (68-62986. Arlene Royer [NARA].)

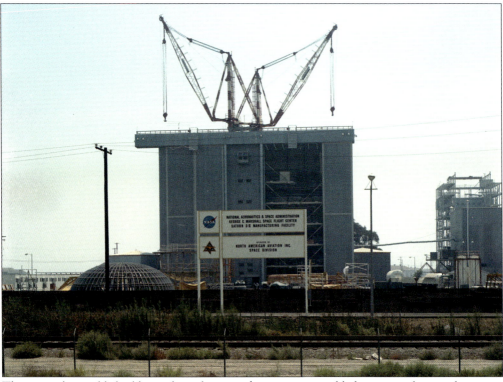

The vertical assembly building, where the second stage was assembled, contained six work stations at which successive major parts of the stage were added. After assembly, the stage was moved to the vertical checkout building, seen in the background in this August 1967 photograph, where some final installations were made and where its systems and components were given final tests. (67-58546. Arlene Royer [NARA].)

The liquid hydrogen tank was formed from six cylinders welded together. Each of the cylinders was assembled from four curved, machined aluminum skins. This image shows the Apollo 13 S-II-8 stage in the vertical assembly building at Seal Beach on November 13, 1967. Unlike the S-IC stage, which had separate tanks, the S-II stage had a common bulkhead that separated the fuel and oxidizer. (67-61113. Arlene Royer [NARA]/Alan Lawrie.)

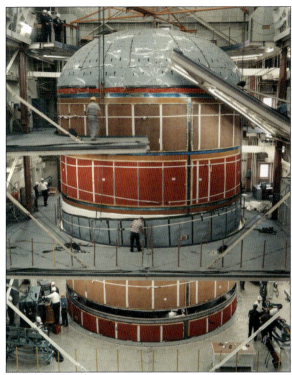

Welding the large cylinders and bulkheads meant that special tooling had to be produced to accurately align the parts of the stage to ensure that the welding was consistent with no porosity or burn through. Here, the pulse arc metal inert gas (MIG) welding of an S-II liquid hydrogen tank cylinder takes place in August 1967. (67-58535. Arlene Royer [NARA].)

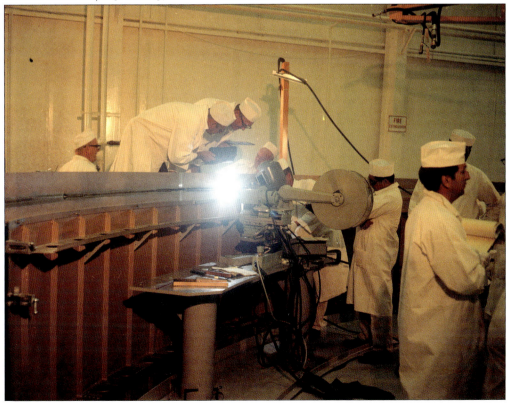

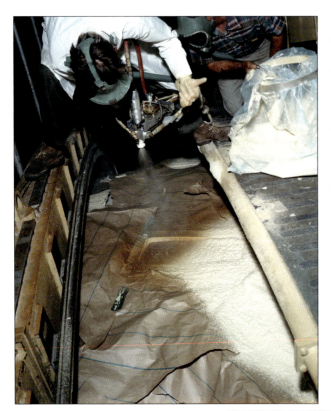

In order to keep the liquid hydrogen fuel at cryogenic temperatures, it was necessary to develop an efficient method to insulate the large surface area of the fuel tank. On the early stages, a foam-filled honeycomb helium-purged system was used. This was later changed to a sprayable polyurethane foam, as seen in this photograph of the foam application on the S-II-8 (Apollo 13) forward bulkhead in August 1967. (67-58528. Arlene Royer [NARA].)

Because the foam was applied to the outside of the tank wall, the metal, being in direct contact with the fuel, was very cold, and the bond was difficult to perfect. Spray foam was first used in some areas on S-II-6 (Apollo 11) but was not fully employed until S-II-8 (Apollo 13). This image shows the foam covering on the S-II-8 forward skirt on September 18, 1967. (67-59784. Arlene Royer [NARA].)

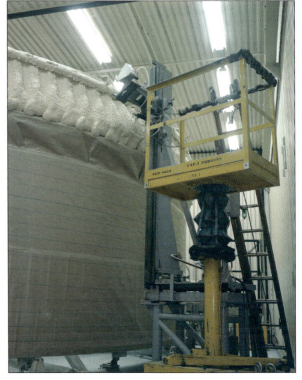

Three structural test stages were produced in order to validate the lightweight structure that was used on the vehicles from S-II-4 (Apollo 9) onwards. One of these stages, S-II-TS-B, was transported by ship from Seal Beach to Port Hueneme, California, and then overland to the Santa Susana Field Laboratory on October 31, 1967, where it underwent structural testing. (67-60426. Arlene Royer [NARA]/Alan Lawrie.)

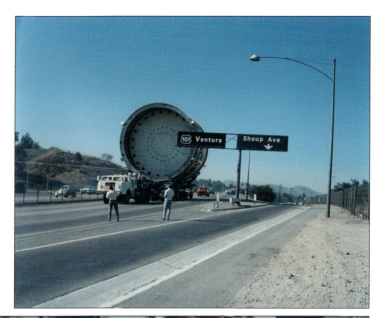

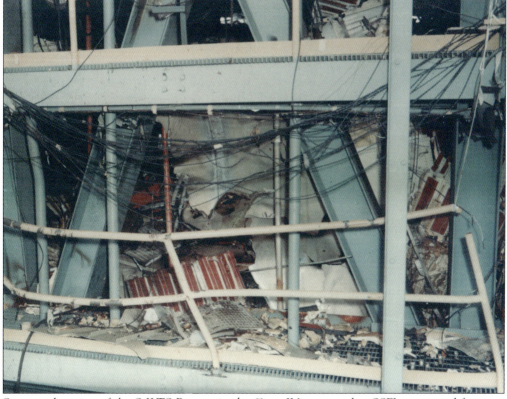

Structural testing of the S-II-TS-B stage in the Coca IV test stand at SSFL continued for more than a year before a catastrophic test failure occurred on December 20, 1968. The explosion, which happened just one day prior to the first manned launch of the Saturn V, was due to mixing hydrogen with oxygen gas that had not been adequately purged. (69-72011. Arlene Royer [NARA]/Alan Lawrie.)

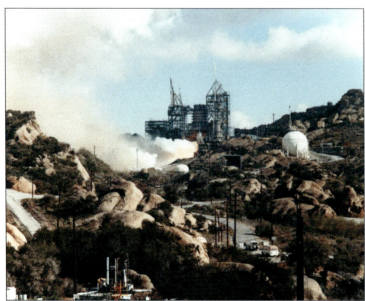

An S-II Battleship stage was used to test the stage propulsion systems in test firings at Rocketdyne's SSFL between 1964 and 1968. Fifty-seven separate firings were performed in this time, with the 352-second firing in the Coca I stand on January 31, 1968, shown here. The rugged terrain of SSFL was also used on occasions by Hollywood film producers to shoot Westerns. (68-63007. Arlene Royer [NARA]/Alan Lawrie.)

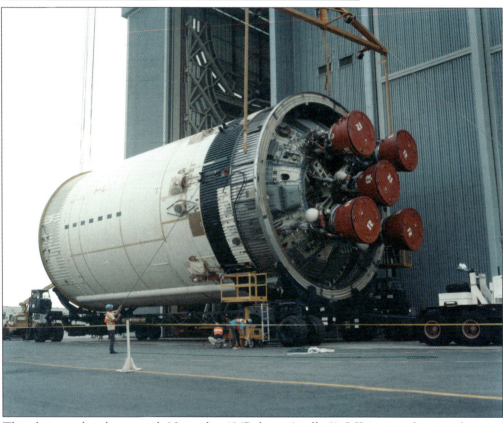

This photograph, taken in early November 1967, shows Apollo 9's S-II-4 stage after completion of systems checkout being moved on a transporter to the Seal Beach vertical assembly building in preparation for shipment to MTF. The five J-2 engines have red transportation covers to protect them. (67-61116. Arlene Royer [NARA]/Alan Lawrie.)

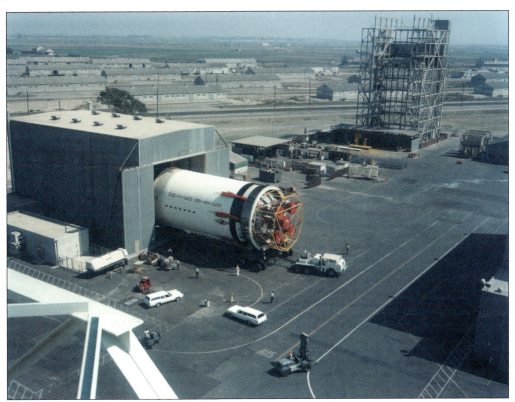

The final Saturn second stage to depart Seal Beach was the S-II-15 stage, which left in September 1970. Here, the S-II-15 stage is being transferred to the Seal Beach Station VII on September 1, 1970, for packaging prior to shipment to MTF. Subsequently, Seal Beach was actually used as a storage facility for several Saturn S-IVB third stages. (70-11853. Arlene Royer [NARA]/Alan Lawrie.)

Because the NAA Seal Beach facility was close to habitation, pressure testing, which potentially could have resulted in explosions, was performed at a remote site, access to which required that the stages be transported along public roads. On March 27, 1969, a convoy of S-II stages is leaving the Seal Beach facility for different locations. S-II-12, at the rear, heads for pressure testing and S-II-9, at the front, to the Seal Beach docks. (69-73499. Arlene Royer [NARA]/Alan Lawrie.)

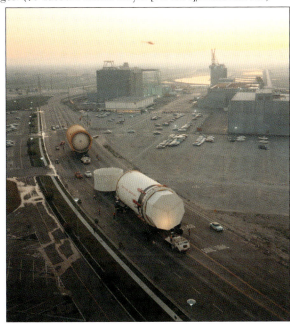

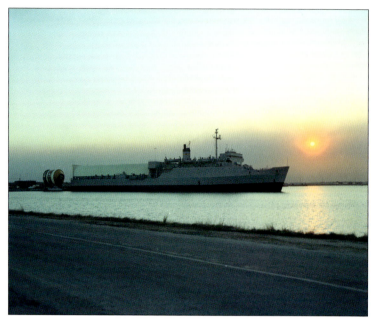

Because of the large distance between the manufacturing location of Seal Beach on the West Coast and the testing location in Mississippi, each stage was transported on the ocean-going ship *Point Barrow* on a journey that passed through the Panama Canal. This photograph shows the *Point Barrow* at the Seal Beach Naval Weapons Station with the S-II-TS-B test stage being loaded on board on October 30, 1967. (67-60412. Arlene Royer [NARA].)

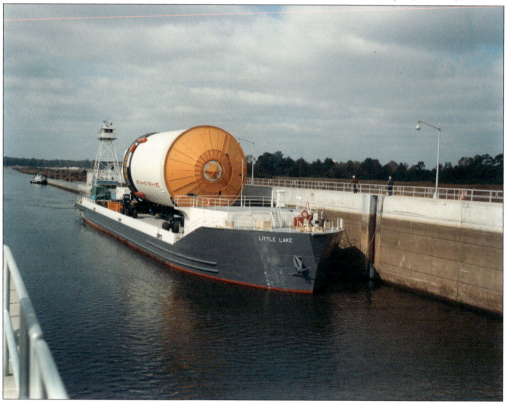

As the *Point Barrow* was too large to go all the way up the Pearl River to MTF, the final part of the journey from Michoud to MTF was completed on one of the open-top barges *Little Lake* and *Pearl River*. This image shows the Apollo 12 S-II-7 stage arriving at MTF on November 12, 1968, on board the barge *Little Lake*. (68-70784. Arlene Royer [NARA]/Alan Lawrie.)

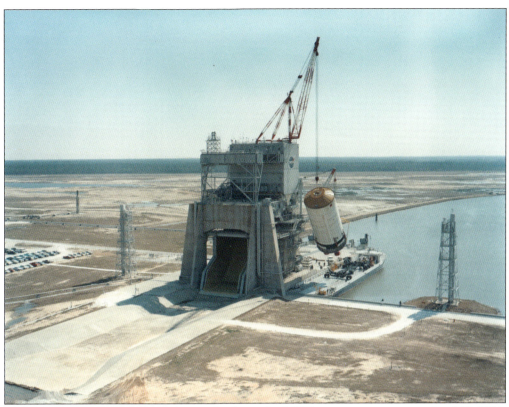

The first flight-type S-II stage to be test fired was the S-II-T test stage, starting on April 23, 1966. However, after eight firings, the stage was destroyed on May 28, 1966, in MTF's A-2 test stand. This was caused by human error due to miscommunication about the test setup. This photograph shows a similar setup with the S-II-5 stage at the A-1 test stand at MTF on March 13, 1968. (68-64160. Arlene Royer [NARA]/Alan Lawrie.)

When the Apollo 9 S-II-4 stage was test fired in MTF's A-2 test stand on January 30, 1968, Tulane University Medical School placed guinea pigs in strategic areas of sound propagation in order to study the acoustic effects of massive sound on the laboratory animals. Thankfully for the animals, the firing was aborted after only 17 seconds. The second firing, on February 10, 1968, is shown in this image. (68-64039. Arlene Royer [NARA]/Alan Lawrie.)

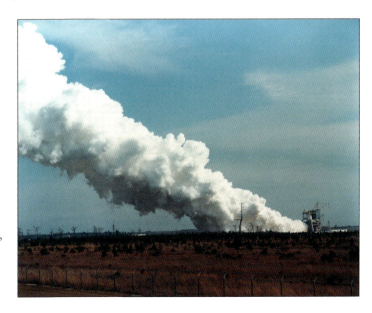

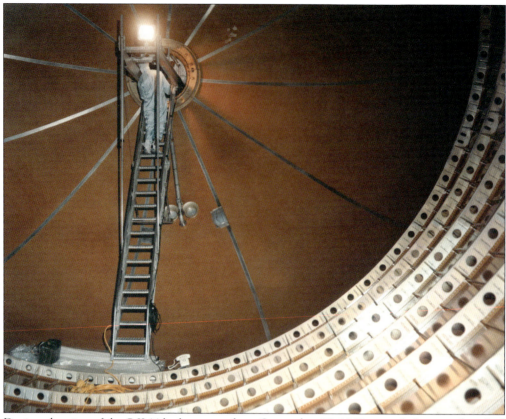

During cleaning of the S-II-14 hydrogen tank on September 30, 1969, a nozzle became detached from a revolving spray mast and fell 40 feet into the inverted stage, causing a small crack in a gore on the forward bulkhead. The repaired area, shown in this January 29, 1970, image inside the hydrogen tank, involved bolting on a structural doubler. Note the telephone at the base of the ladder. (70-04762. Arlene Royer [NARA]/Alan Lawrie.)

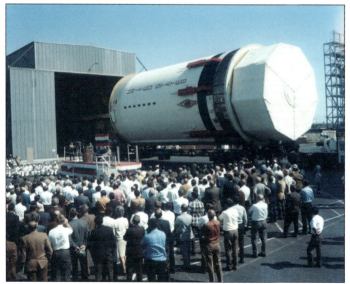

The 15th and final S-II second stage to be manufactured and assembled at Seal Beach was given a special farewell ceremony on September 8, 1970, prior to its departure to MTF for static fire testing. This stage can now be seen on display at the Johnson Space Flight Center in Houston. (70-11869. Arlene Royer [NARA]/Alan Lawrie.)

Three

SATURN V THIRD STAGE

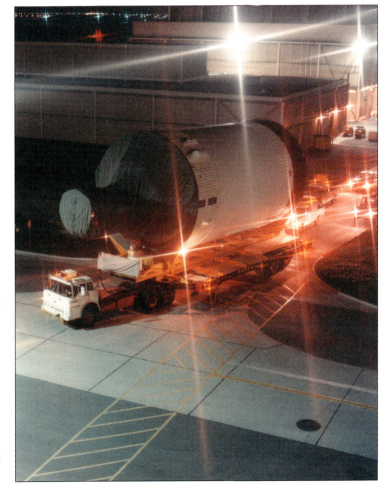

The Saturn V third stage was designated S-IVB and was assembled by the Douglas Aircraft Company (later called McDonnell Douglas) at its site at Huntington Beach, California. In this image, the Apollo 16 S-IVB-511 stage is leaving the site on September 16, 1969, heading for Douglas's Sacramento Test Operations static firing site. (69-02593. Arlene Royer [NARA]/ Alan Lawrie.)

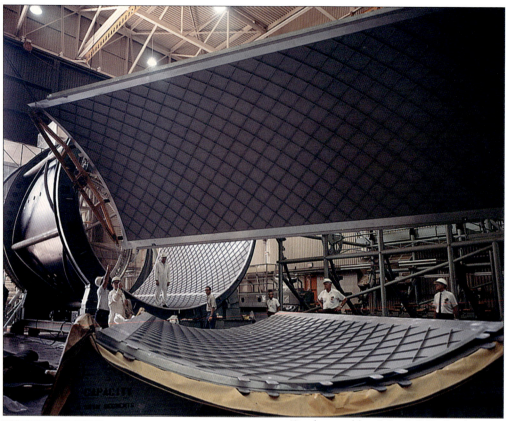

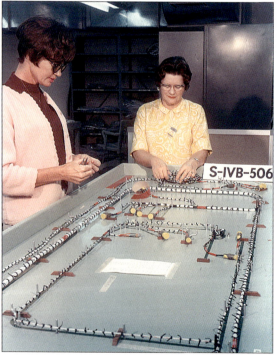

Final assembly of the S-IVB third stages took place at the Douglas Huntington Beach facility, with major sub-assemblies arriving from DAC's Santa Monica plant. The cylindrical section of the stage was the liquid hydrogen tank, assembled from seven curved panels welded together, and the liquid oxygen tank was attached at the bottom via a common bulkhead. The Apollo 15 S-IVB-510 hydrogen tank panels are shown being assembled in July 1967. (Phil Broad/Mike Jetzer.)

Manufacture, assembly, and testing of each stage involved many people at different locations with specialist skills. In this mid-1967 photograph, technicians at Huntington Beach are assembling the electrical harness to be installed on the Apollo 11 S-IVB-506N stage. Four of the S-IVB stages were given an "N" (for new) designation after the loss of the S-IVB-503 stage in an explosion. (Phil Broad/Mike Jetzer.)

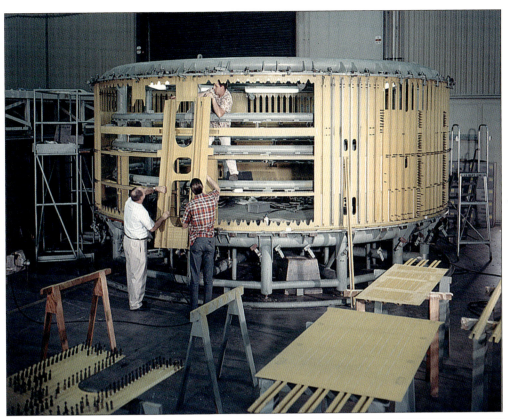

The aft skirt assembly provided the interface between the third stage propellant tank assembly and the aft inter-stage assembly. It was a cylindrical aluminum structure comprised of curved panels stiffened with stringers. Two solid-propellant ullage motors and two liquid auxiliary propulsion system engine modules were attached to the aft skirt assembly. The mid-1967 Apollo 12 S-IVB-507 assembly is shown here. (Phil Broad/Mike Jetzer.)

The hydrogen tank forward dome and the oxygen tank were manufactured at Santa Monica and transported to Huntington Beach for stage assembly. This photograph shows the welding of the Apollo 11 S-IVB-506N forward dome to the hydrogen tank cylinder in April 1967. The forward dome had been formed by the welding together of nine gores or "orange peel" segments. (Phil Broad/Mike Jetzer.)

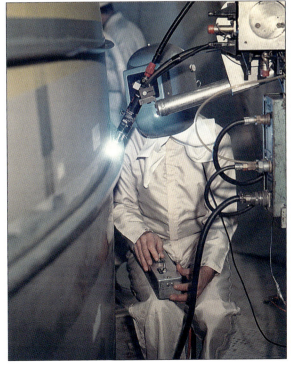

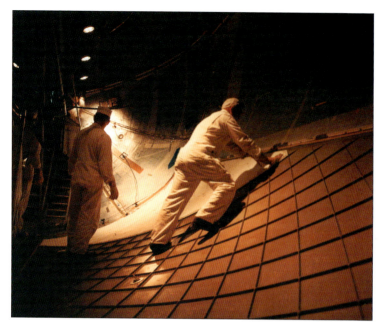

Douglas's preferred solution to the problem of keeping the liquid hydrogen cold was to insulate the inside of the tank wall rather than the outside, as done on the second stage. This allowed for a warmer and hence more reliable sealing surface between the tiles and the tank wall. Repair work on the insulation tiles takes place on the S-IVB-507 stage in August 1967. (67-58365. Arlene Royer [NARA]/Alan Lawrie.)

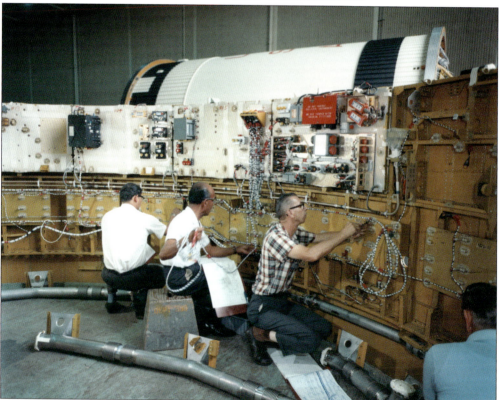

The Huntington Beach site was the headquarters for the Douglas Missile and Space Systems Division. In this image, technicians install wiring harnesses on the Apollo 11 S-IVB-506N aft skirt in August 1967. In 1967, Douglas became McDonnell Douglas, and the Huntington Beach plant is now owned by Boeing. (67-58368. Arlene Royer [NARA]/Alan Lawrie.)

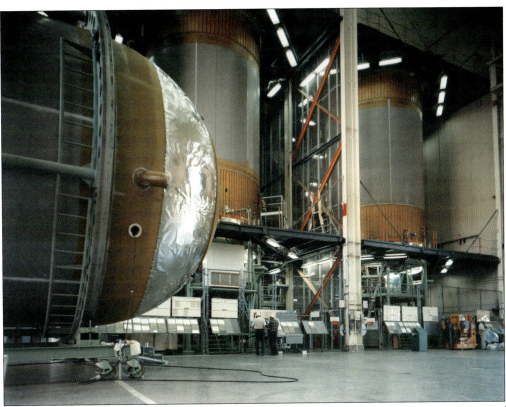

Eight tower positions were available at Huntington Beach for vertical assembly and checkout of completed vehicles. Two vertical towers provided for the final factory tests on finished third stages prior to shipment from the plant for test firing. In this August 1967 image, from left to right are S-IVB-507, S-IVB-506N, and S-IVB-212. The last of these became the Skylab Orbital Workshop (OWS) and was launched on the final Saturn V. (67-58373. Arlene Royer [NARA]/Alan Lawrie.)

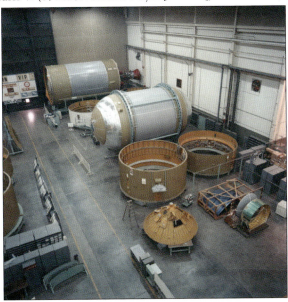

This photograph, taken in March 1970, shows the assembly area at Huntington Beach as some of the final Saturn hardware was processed. At top left is the S-IVB-513 stage, complete with J-2 engine. This stage is now on display at the Johnson Space Center (JSC) in Houston, Texas. At center is the S-IVB-515 stage tank assembly, which became the back-up Skylab OWS. In the foreground are parts of the S-IVB-212 stage. (70-06056. Arlene Royer [NARA]/Alan Lawrie.)

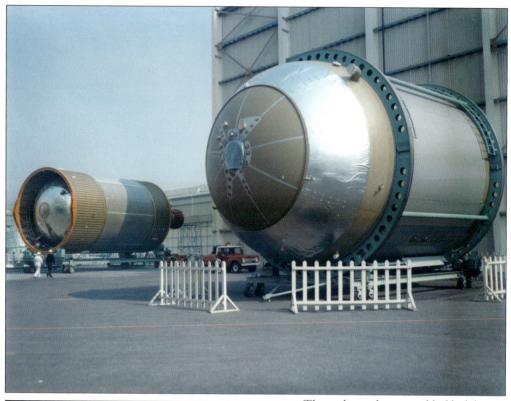

Throughout the second half of the 1960s and beyond, Huntington Beach handled a large number of Saturn S-IVB stages at various levels of build and test. In this photograph, taken in February 1970, the S-IVB-513 stage in the background is en route to the Manufacturing Building where it will be painted, while in the foreground is the S-IVB-515 tank assembly without forward skirt, aft skirt, or engine attached. (70-05608. Arlene Royer [NARA]/Alan Lawrie.)

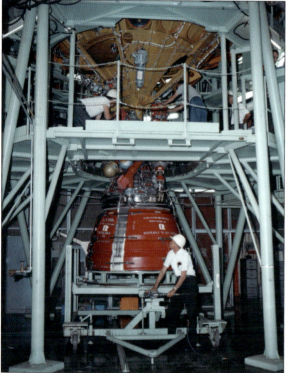

One of the final stages of assembly was to install the single J-2 engine that was delivered from Rocketdyne, located just across Los Angeles. This photograph shows the Apollo 11 S-IVB-506N J-2, serial number J-2101, being mated to the stage by technicians in Assembly Tower No. 5 at Huntington Beach on August 21, 1967. (67-58379. Arlene Royer [NARA]/Alan Lawrie.)

Shown here is an exterior view of the vertical checkout towers at Huntington Beach, which at the time of the photograph in July 1967 contained the Apollo 10 S-IVB-505N stage on the left (position number seven) and the Saturn IB stage S-IVB-211 on the right (position number eight.) The S-IVB-211 stage never flew and is now part of an outdoor Skylab display at the USSRC in Huntsville. In 1965, the assembly towers had been moved here from Santa Monica, where they were used on the S-IV program. (Phil Broad/Mike Jetzer.)

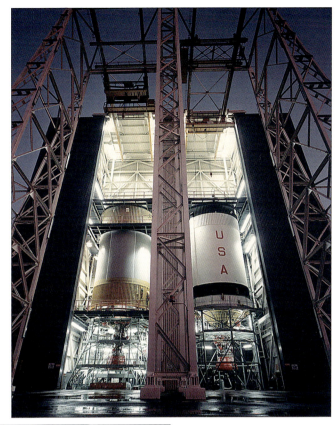

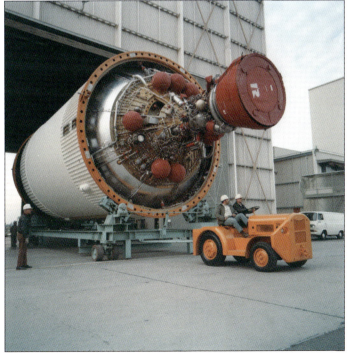

As the Saturn production contracts were reaching completion, there was a surplus of hardware. This stage, S-IVB-512, seen here being moved to the Manufacturing Building at Huntington Beach in January 1970, remained in storage for the next year. It eventually sent the last crew, Apollo 17, to the moon, and it has the distinction of being the only Saturn V stage to be launched having never been test fired on the ground. (70-04827. Arlene Royer [NARA]/Alan Lawrie.)

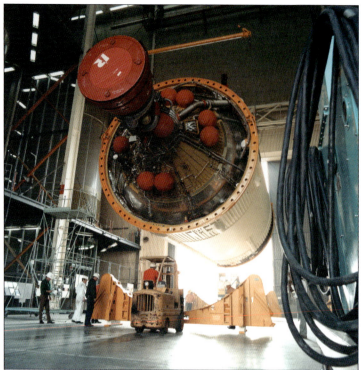

One of the final activities to be performed on each stage at Huntington Beach was to measure the weight and center of gravity. Here, the Apollo 14 S-IVB-509 stage is being prepared for "weight and balance" in March 1969. Each stage was then flown by Super Guppy aircraft from Los Alamitos Naval Air Station to Mather Air Force Base for static fire testing at SACTO. (69-73476. Arlene Royer [NARA]/ Alan Lawrie.)

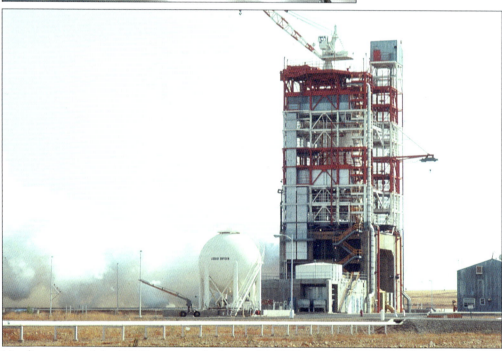

Hot fire testing of the S-IVB stages took place at SACTO, near Sacramento, where Douglas had two test stands (Beta I and Beta III). This image shows the firing of a Saturn S-IVB stage in one of the Beta test stands. Full details of the SACTO activities can be found in the Images of Modern America book *Sacramento's Moon Rockets*. (Dick Serrano.)

During the preparations for acceptance test firing of the S-IVB-503 stage in the Beta III test stand on January 20, 1967, there was an explosion that destroyed the stage. It would have flown on Apollo 8, sending the first crew around the moon. It was the only Saturn V stage planned for flight that was destroyed in ground testing. Failure was due to human error at a lower-tier subcontractor. (G83-84. Jim Porter/Terri Pennello.)

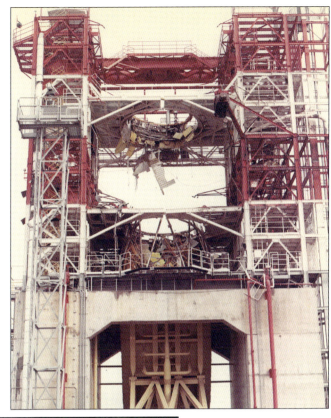

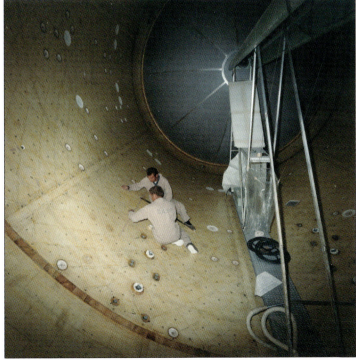

The final Saturn V to be launched substituted the third stage for the Skylab space station, which was launched into Earth orbit. The Skylab OWS was a modified Saturn IB second stage, S-IVB-212. Originally built as a launcher stage, the S-IVB-212 was converted to a habitable module. The three Skylab crews were to live in its hydrogen tank. During May 1969, crews performed the conversion of the interior of the stage at Huntington Beach. (69-75845. Arlene Royer [NARA]/Alan Lawrie.)

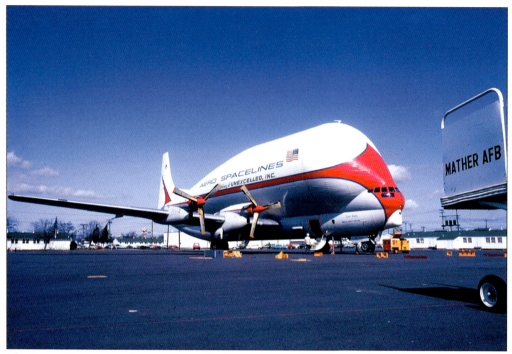

The Super Guppy aircraft was purpose-built to carry the S-IVB stages from Huntington Beach to Sacramento and then onwards to the Kennedy Space Center in Florida following the completion of the static firing and other mechanical and electrical tests. Mather Air Force Base was the closest airfield to SACTO and so was in frequent use carrying S-IVB stages, as seen in this April 1967 image. (Dick Serrano.)

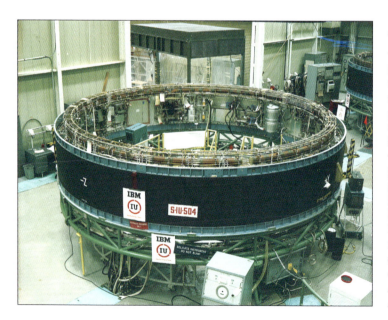

The Instrument Unit was a three-foot-high ring attached to the top of the S-IVB third stage. It provided the guidance and control for the complete Saturn V and sent all the instrumentation parameters to the ground. It was assembled by IBM at its facility in Huntsville's Research Park. This image shows the Apollo 9 S-IU-504 during sub component checkout at IBM in August 1967. (67-58415. Arlene Royer [NARA].)

Four

F-1 AND J-2 ENGINES

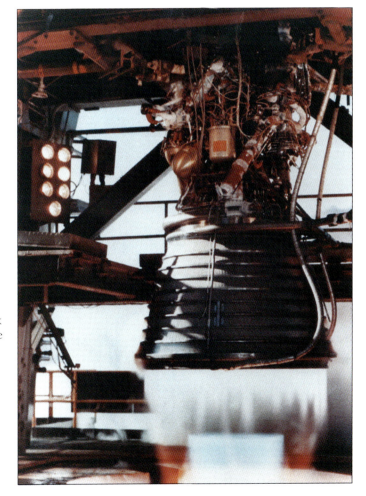

Five J-2 engines were used on the Saturn V second stage, and one was used on the third stage. They were manufactured by Rocketdyne at Canoga Park in Los Angeles. Acceptance testing took place at Rocketdyne's Santa Susana Field Laboratory, and this photograph shows the final ever test firing of a J-2, J-2152, on January 2, 1970. This engine was destined for the S-II-15 stage. (70-04809. Arlene Royer [NARA]/Alan Lawrie.)

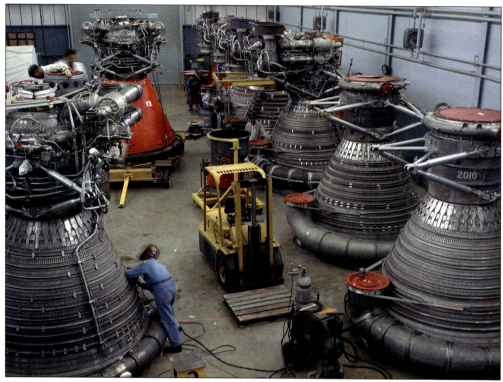

The F-1 engines were the largest rocket engines ever produced, burning liquid oxygen and RP-1 kerosene. There were 98 flight-standard engines produced, the first 28 having a thrust of 1.5 million pounds and the remaining engines being uprated to 1.522 million pounds. This photograph shows around nine F-1 engines in the Engine Preparation Shop at Canoga Park on March 19, 1965. (65-21185. NASA.)

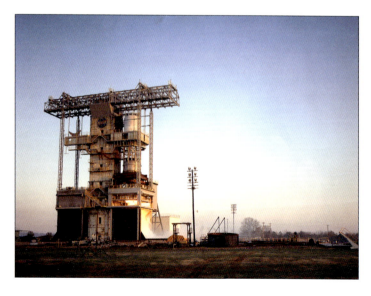

Although most engine testing was performed by Rocketdyne, some engines were delivered to NASA and tested at MSFC. Between December 1963 and September 1966, seventy-five test firings of single F-1 engines took place in the S-I/IB static test tower in the East Test Area at MSFC. This 36-second firing of engine F-1001 took place at 4:15 p.m. on December 17, 1963. (63-35341. NASA.)

In 1965, a new stand was built to test-fire single F-1 engines in the West Test Area at MSFC. Between July 1965 and February 1969, there were 107 engine firings. This photograph shows a firing of engine F-5038-1 as viewed from the top of the S-IC test stand in May 1968. This test stand was demolished in November 2012. (68-66986. NASA.)

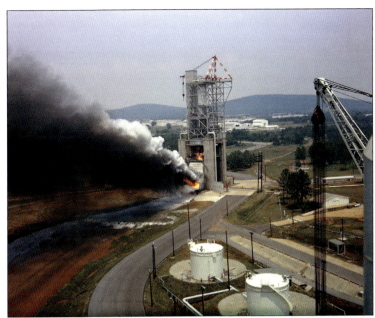

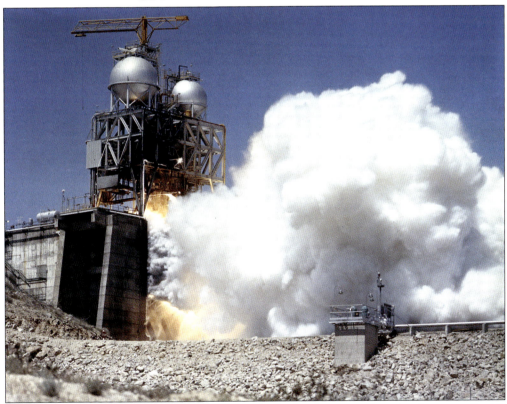

Because of the high thrust level, it was necessary for Rocketdyne to perform hot fire tests of its F-1 engines at a facility remote from its Los Angeles manufacturing and testing sites. Consequently, all F-1 engines underwent acceptance testing at the AFRPL at Edwards Air Force Base. This image shows a firing in one of the five test stands used for F-1 testing at the AFRPL. (64-01638. NASA.)

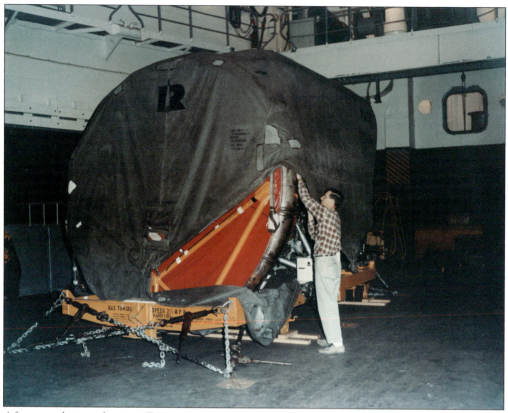

After completion of testing, F-1 engines were transported to Michoud by a variety of methods over the years. Initially, they were flown by the Pregnant Guppy aircraft. Later, they were transported across the country on a truck. The last few engines hitched rides on *Point Barrow*, which was carrying S-II stages from Seal Beach to Michoud. Here, the F-6073 engine arrives at Michoud on February 16, 1968, after becoming the first engine to travel by ship. (68-64127. Arlene Royer [NARA]/Alan Lawrie.)

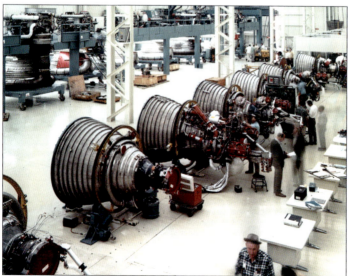

This March 9, 1966, image shows the engine assembly area at Rocketdyne's Canoga Park, with J-2 engines being worked on in the foreground and F-1 engines lined up at the back. Rocketdyne opened the 51-acre facility in 1955 and designed and manufactured all the Saturn V rocket engines. The Canoga Park site closed in March 2014. (Vince Wheelock.)

In addition to full engine testing, component testing took place throughout the 1960s in order to refine performance and overcome any problems that occurred. Here, Rocketdyne performs J-2 injector test firings in one of the Vertical Test Stands in the Bowl Area of SSFL in October 1967. (68-62882. Arlene Royer [NARA]/Alan Lawrie.)

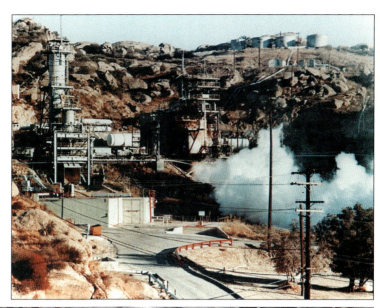

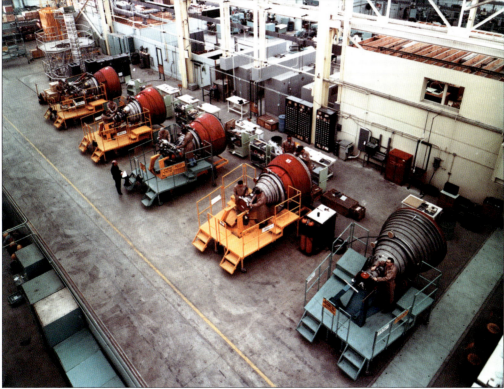

This image shows the J-2 rocket engine assembly line at Canoga Park on January 28, 1964. Initially, manufacturing of components and final assembly of both Saturn engines was carried out in eight buildings in the Canoga Park complex. As well as machining facilities, there were two large brazing furnaces for the brazing of the thrust chamber tubes and injectors. Special areas for precision cleaning and assembly and for various types of nondestructive inspection existed. (Vince Wheelock.)

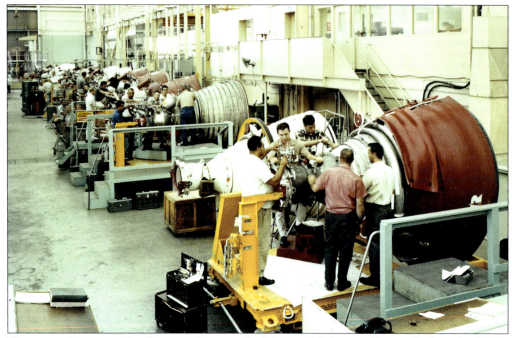

This is another view of technicians busy on the J-2 assembly line at Canoga Park in the 1960s. The first 59 J-2 engines had a thrust of 200,000 pounds, after which the thrust was uprated to 230,000 pounds. In addition to their usage on the Saturn V second and third stages, J-2 engines were used on the S-IVB Saturn IB second stages. (NASA.)

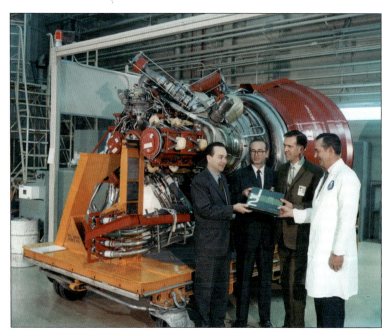

Rocketdyne produced 152 flight-standard J-2 engines. This photograph shows the handover of the final engine, J-2152, and its log book to NASA at Rocketdyne's Canoga Park on January 28, 1970. (70-06065. Arlene Royer [NARA]/ Alan Lawrie.)

Five

BEFORE THE
MOON LANDINGS

The three Saturn V stages plus the IU and the Apollo spacecraft were all transported to the Kennedy Space Center, where they were assembled into a complete Saturn V rocket in the VAB. Two launchpads (39A and 39B) were built for the Saturn V rockets. This image shows the first flight vehicle, Apollo 4, or SA-501, on Launchpad 39A the night before its launch on November 9, 1967. (AS4-S67-044. J.L. Pickering.)

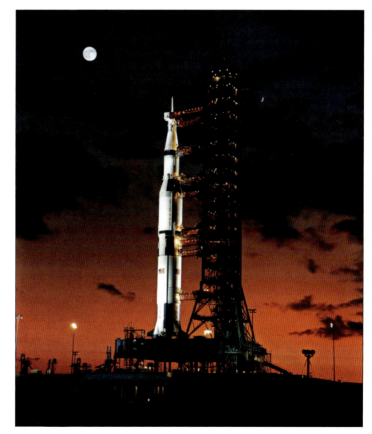

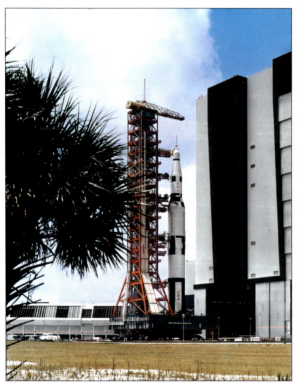

The first complete Saturn V to be assembled was the Facilities Checkout Vehicle, SA-500F, which rolled out from High Bay 1 to pad 39A on May 25, 1966. In June, the vehicle was rolled back to the VAB for two days because of the approach of Hurricane Alma. SA-500F spent the summer at the pad verifying interfaces and allowing propellant loading tests to take place. (KSC-66PC-75. J.L. Pickering.)

The SA-500F rocket is seen here at the pad on July 22, 1966. After testing was finished, the vehicle was rolled back into the VAB on October 14, 1966. Before destacking took place, the engineers decided to perform an unofficial rocking test of the tall rocket in order to identify its natural swaying frequency. One group of technicians pushed with their feet to get the rocket moving. (KSC-66C-6962. J.L. Pickering.)

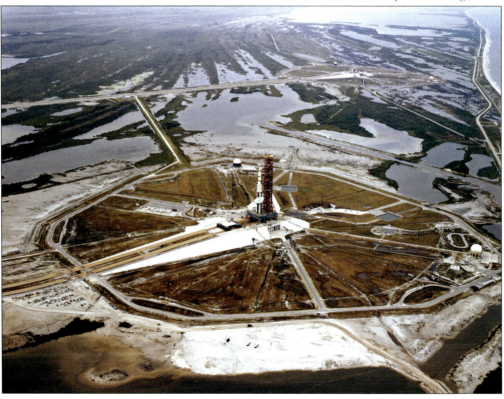

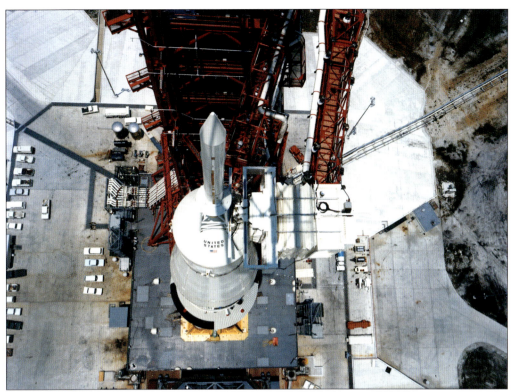

The Launch Escape System (LES) tower at the top of the rocket is seen in this image. A second group of technicians tied a rope around the LES and tugged in time with the men on the other side pushing. The rocket swayed, yielding valuable data, but they pulled too hard, and the LES came right off the rocket. The incident was kept quiet until film emerged in 2008, together with the recollections of those involved. (KSC-66PC-5431. J.L. Pickering.)

In order to measure the structural response of the full Saturn V rocket, a Dynamic Test Vehicle, SA-500D, was assembled at MSFC in Huntsville. The Saturn V comprised all three stages and Apollo payload, which were constructed to be structurally representative but not functional from a propulsion standpoint. Assembly of the complete vehicle in the purpose-built Dynamic Test Stand at MSFC was completed on December 3, 1966, and the full vehicle was tested between January and March 1967. (USSRC.)

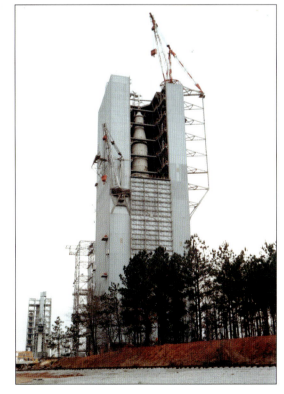

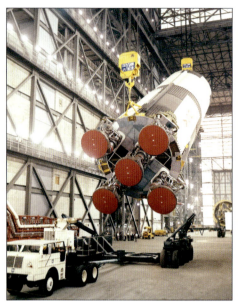

The first Saturn V to be launched was the unmanned Apollo 4, SA-501. The S-IC-1 stage was shipped on the barge *Poseidon* from MSFC in Huntsville to KSC, although it was delayed in New Orleans due to a strike between the towing subcontractor and the union. It eventually arrived at KSC on September 12, 1966, and is seen here being processed in the VAB on October 20, 1966. (AS4-66C-8879. J.L. Pickering.)

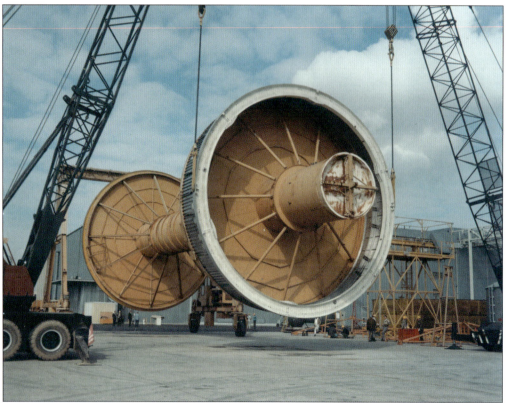

The S-II-1 second stage for Apollo 4 was three months behind the first stage in arriving at KSC. Consequently, to allow testing of an assembled Saturn V to take place, the first flight Saturn V was initially assembled with an S-II stage simulator acting as a spacer between the first and second stages. The simulator is shown here in later storage at Seal Beach on March 13, 1970. (70-06027. Arlene Royer [NARA]/Alan Lawrie.)

The VAB had four High Bays (HB), with bays 1 and 3 facing the launchpads (east) and bays 2 and 4 facing west. The Apollo 4 Saturn V was assembled in HB1 on Mobile Launcher number one (ML-1). After the first stage was erected on October 27, 1966, it was another 10 months before the rocket was rolled out to the launchpad, with the second stage being de-erected once and the third stage twice for various checks. (67-54387. J.L. Pickering.)

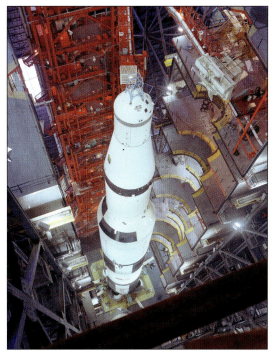

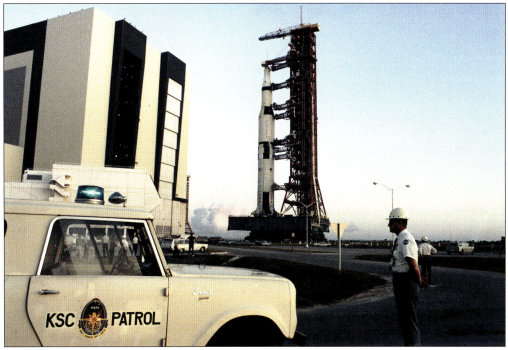

Roll out of Apollo 4 to Launchpad 39A occurred on August 26, 1967. Each Saturn V rocket was transported from the VAB to the launchpad on one of three Mobile Launchers, each of which comprised a Launcher Base and Launch Umbilical Tower (LUT), which included a hammerhead crane on the top. There were two Crawler Transporters that were used to carry the Mobile Launcher and Saturn V rocket to the launchpad. (AS4-67HC-542. J.L. Pickering.)

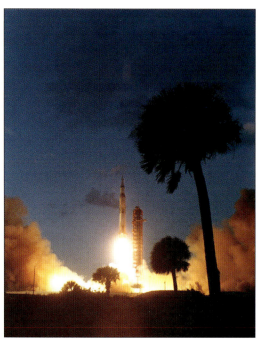

The first launch of a Saturn V occurred at 7:00 a.m. Eastern Standard Time on November 9, 1967. The unmanned Apollo 4 lifted off with a combined engine thrust of 7.5 million pounds. The launch was very successful, with all stages performing as planned, leading to a successful splashdown of the Command Module 8 hours and 37 minutes after launch. The mission included the first reignition of an S-IVB stage in flight. (AS4-67HC-734. J.L. Pickering.)

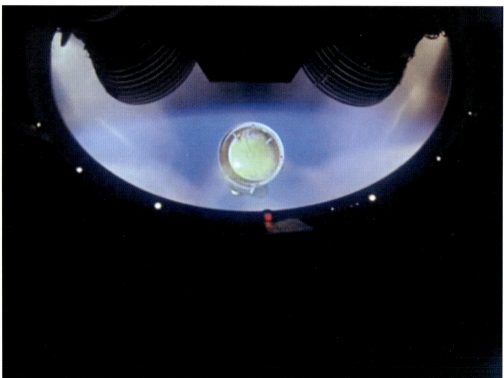

During the launch of Apollo 4, two film camera pods were mounted at the base of the S-II stage to film the separation of the S-IC stage and the inter-stage ring. Both pods were later ejected and recovered in the sea. This still from the film shows the S-IC first stage falling away, with the J-2 engines of the S-II stage in the foreground at the top. (NASA/Mike Jetzer.)

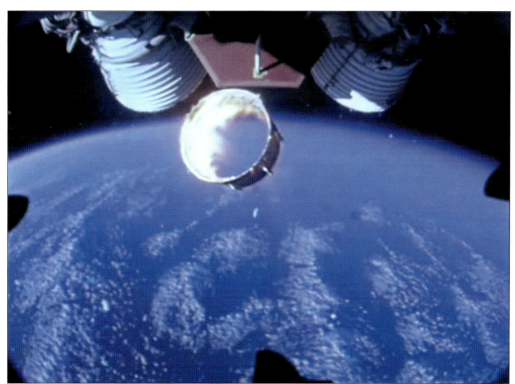

Later in the same film sequence, the Apollo 4 inter-stage ring can be seen separating from the S-II second stage. At this time, the J-2 engines are firing, but their exhaust plumes are invisible. On the later Apollo 6 launch, there were six film camera pods and two television cameras on the first two stages. However, only two of the camera pods were recovered. (NASA/Mike Jetzer.)

The second unmanned flight of the Saturn V was designated SA-502 or Apollo 6. Again the S-II second stage was delivered late, necessitating the build-up of the rocket with the S-II stage simulator for three months between March and June 1967. Eventually the S-II-2 stage arrived, and it was attached to the first stage in the VAB on July 13, 1967. (AS6-68HC-128. J.L. Pickering.)

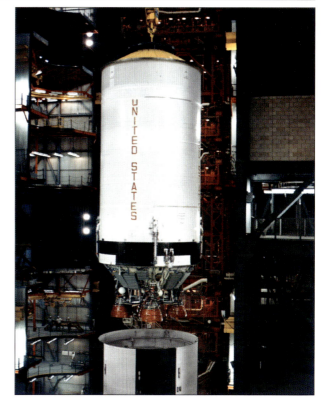

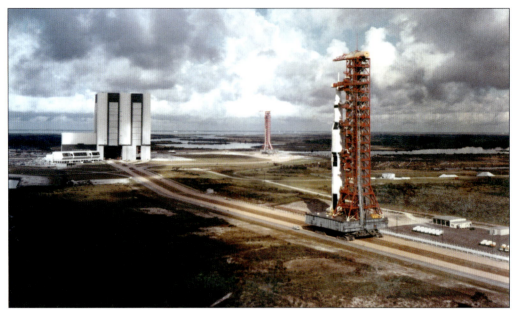

The Apollo 6 vehicle was transported the three miles from the VAB to Launchpad 39A on February 6, 1968. The rocket had been assembled on the second Mobile Launcher, ML-2. With the Apollo 8 rocket being assembled on ML-1 in the VAB at this time, the third Mobile Launcher, ML-3, can be seen parked beside the VAB. (J.L. Pickering.)

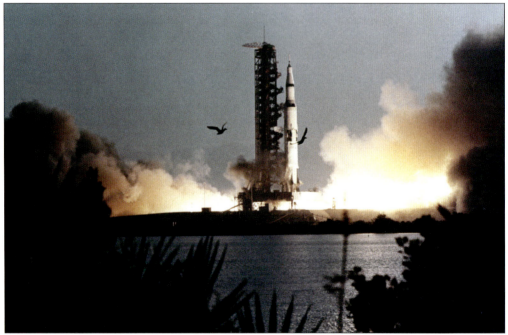

Launch of SA-502, Apollo 6, took place at 7:00 a.m. Eastern Standard Time on April 4, 1968. For 30 seconds during the first-stage section of flight, longitudinal oscillation, known as "pogo," was experienced. This was due to a coupling of the stage structure natural frequency to an excitation of the F-1 engines linked to an oxidizer line frequency. The solution, implemented on the next flight, was to add in a helium-filled shock absorber. (AS6-68PC-57. J.L. Pickering.)

Here is a view of the Apollo 6 launch taken from an automatic camera on the LUT. During second stage firing, engine number two shut down after 412 seconds, and the IU, in attempting to complete the shutdown, also shut down a good engine (number three) two seconds later due to incorrect wiring. The problem on engine number two reoccurred on the third-stage engine, preventing reignition. The problem was caused by fractured propellant lines. (AS6-68HC-201. J.L. Pickering.)

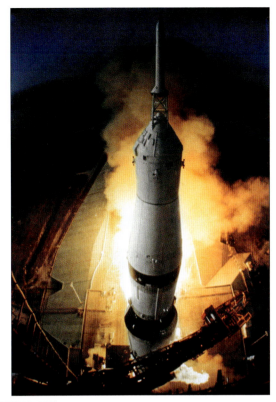

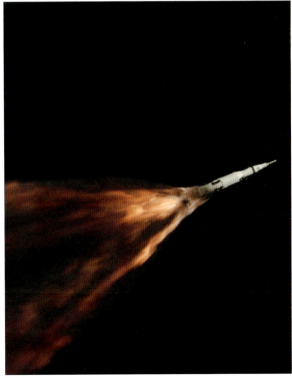

Sections of the Spacecraft Lunar Module Adapter (SLA) panels near the top of the vehicle fell off after 133 seconds, as can be seen as two black smudges on the SLA's truncated cone in this tracking camera image. The cause was pressure build-up in the honeycomb panels as the ambient pressure decreased with altitude. (AS6-S68-29735. J.L. Pickering.)

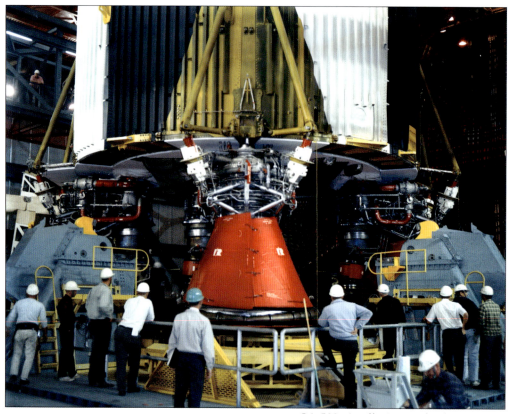

SA-503, Apollo 8, was the first manned Saturn V flight and the first time the Saturn V would send a capsule to orbit the moon. The first stage was erected on ML-1 on December 30, 1967. A problem with engine number one on the S-IC first stage resulted in the need for its replacement. It was unique in the Saturn V program for an engine to be replaced on the launch vehicle after delivery to KSC. (AS8-68HC-71. J.L. Pickering.)

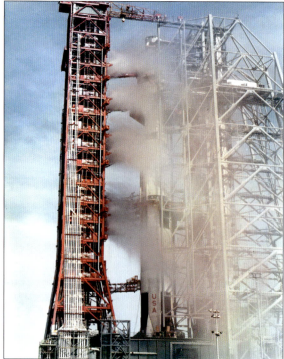

Shown here is a test of the Water Control System, used for cooling and quenching the launch site and storage areas before, during, and after a launch. This test, with the Apollo 8 launch vehicle on pad 39A, took place on November 12, 1968, just over a month before launch. (J.L. Pickering.)

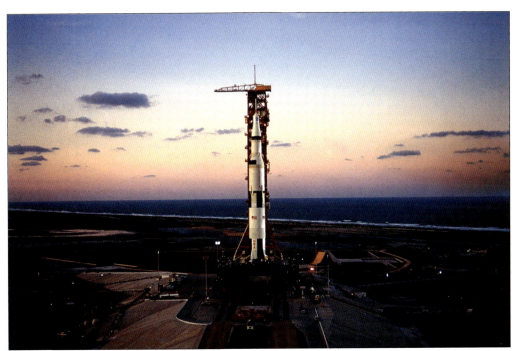

The Apollo 8 rocket stands on Launchpad 39A in the days leading to launch. The Mobile Service Structure (MSS), which provided access for technicians, was removed from the Saturn V at this time. The flame deflector can be seen in its standby position and would be moved underneath the rocket for launch. (AS8-S68-55416. J.L. Pickering.)

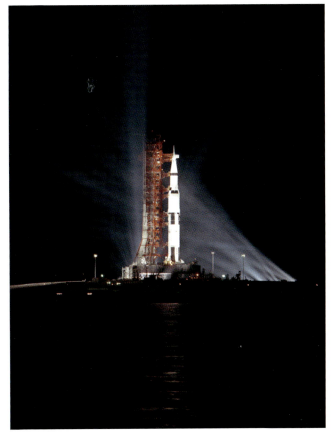

Apollo 8 is pictured under arc lights on the morning of launch, December 21, 1968. The S-II stage had been assembled on the rocket stack since January 31, 1968. However, as a result of the mission of Apollo 8 being upgraded to being manned, it was necessary to uprate the S-II stage inspection criteria and perform a cryogenic proof test. Therefore, between April and July 1968, it was back at MTF for the additional test, which was completed successfully. (J.L. Pickering.)

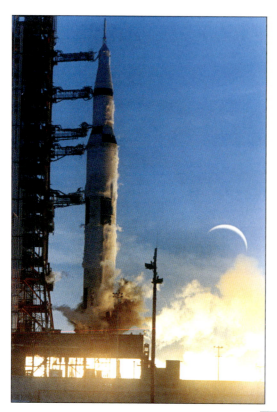

The first manned flight of the Saturn V occurred at 7:51 a.m. Eastern Standard Time on December 21, 1968, when Apollo 8 launched astronauts Frank Borman, James Lovell, and Bill Anders on a mission that took them around the moon 10 times. Famously, the crew read Genesis 1: 1–10 from lunar orbit on Christmas Eve. (AS8-KSC-68PC-329. J.L. Pickering.)

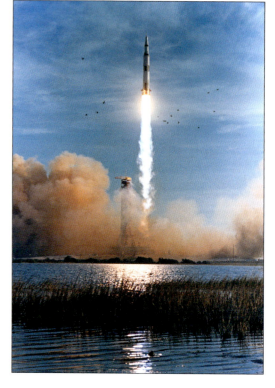

The J-2 engines on the second and third stages incorporated flexible Augmented Spark Igniter (ASI) lines that overcame the problem of fractured propellant lines, which had caused engines to shut down on the previous flight. S-II engine number five experienced a three-percent thrust reduction, and high-amplitude oscillations at 18 hertz were evident during the latter part of the second stage firing. These were determined to be LOX pump–induced, and later corrections were implemented. (AS8-S68-56050. J.L. Pickering.)

The Apollo 8 launch occurred just half an hour after sunrise. As the rocket climbed in the sky, the first stage exhaust became illuminated by the sun and the rocket appeared to head down over the horizon, as viewed from the ground in this photograph. Despite the failure of a structural test item the day before (see page 25), the launch was very successful. (J.L. Pickering.)

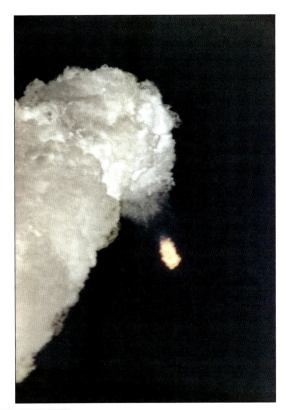

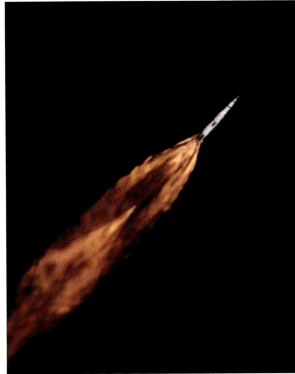

Apollo 8 was the final mission to include on-board cameras to record the rocket in flight. As well as two television cameras with live images of the engine bay, there were four recoverable film cameras on the S-IC stage, although only one was successfully recovered. This showed the propellant in the LOX tank by viewing into the tank through a fiber-optic bundle. (AS8-68PC-336. J.L. Pickering.)

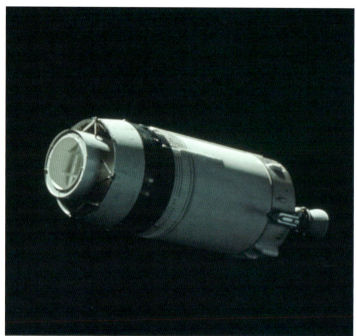

The Apollo 8 S-IVB third stage was the first to successfully perform a Trans-Lunar Injection (TLI) firing with a manned spacecraft. This involved a second firing of the single J-2 engine following repressurization of the hydrogen propellant tank. After separation of the Apollo spacecraft, the crew was able to turn around and capture images of the S-IVB-503N stage (together with its Lunar Module Test Article attached) as it started its indefinite life in heliocentric orbit. (AS08-16-2584. NASA.)

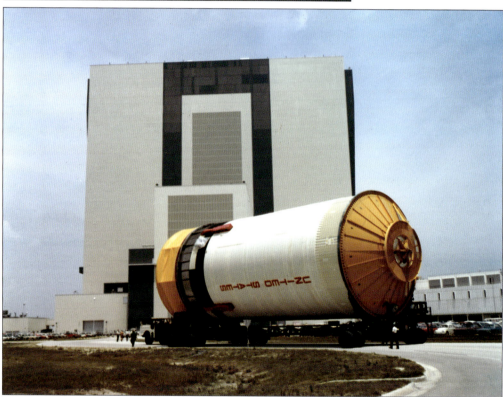

The Apollo 9 second stage, S-II-4, arrived at KSC on May 15, 1968, having spent five days on the barge Orion traveling from MTF via the Intracoastal Waterway. This was the first of the S-II stages to include lightweight tanks and higher-thrust J-2 engines. (AS9-68HC-273. J.L. Pickering.)

Rollout of the Apollo 9 SA-504 vehicle from the VAB HB3 to Launchpad 39A occurred on January 3, 1969. Apollo 9 had been erected on ML-2 in the VAB. This view, looking up from the ground, clearly shows the swing arms on the LUT that provided service access to the launch vehicle right up until launch. Four of the arms retracted prior to ignition, and five rotated away as the rocket left the pad. (AS9-116KSC-369C. J.L. Pickering.)

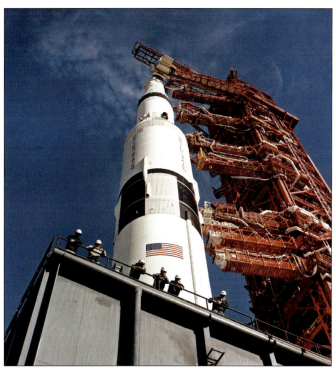

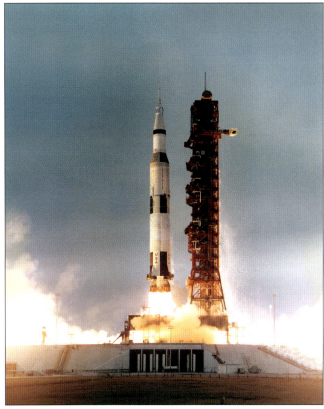

The launch of Apollo 9, SA-504, took place at 11:00 a.m. Eastern Standard Time on March 3, 1969. The crew of Jim McDivitt, Rusty Schweickart, and Dave Scott tested the Lunar Module (LM) in Earth orbit and performed the first Apollo spacewalk. It was the only manned Saturn V mission in which the crew remained in Earth orbit. (AS9-69C-1493. J.L. Pickering.)

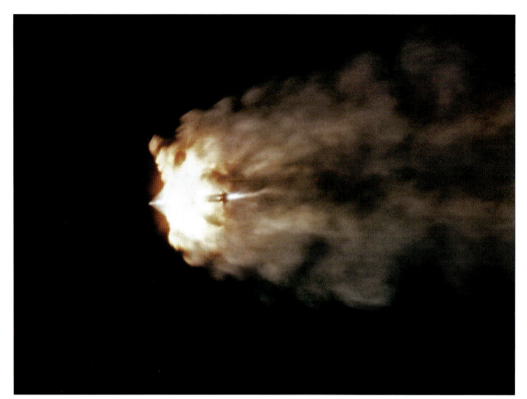

In this photograph, taken by a tracking camera, the Apollo 9 first stage has just separated from the second stage. The fireball was caused by a combination of the eight first stage retro-rocket motors (firing for 0.67 seconds) pushing the spent first stage away and the eight second stage ullage motors (firing for 4 seconds) settling the propellants over the engines and pushing the rocket upwards. (AS9-69PC-67. J.L. Pickering.)

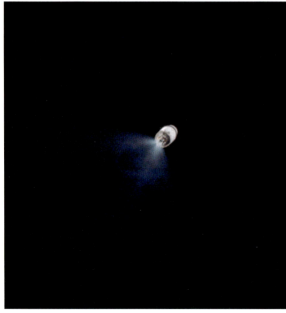

Uniquely in the Saturn V program, it was possible for the Apollo 9 crew to witness a firing of the S-IVB stage. This was because the S-IVB was not used to take the crew to the moon. The S-IVB was fired a second time for 62 seconds, and the start of this burn was photographed by the crew. A later third burn, performed as an experiment, exhibited three-percent thrust reduction due to engine thrust chamber pressure oscillations. (AS09-19-2953. NASA.)

Apollo 10, SA-505, spacecraft erection in the VAB was completed by December 30, 1968. This photograph shows the attachment of the S-IVB third stage on top of the S-II second stage. The Apollo 10 vehicle was the first to be stacked on ML-3 and the first to be assembled in HB2 on the west side of the VAB. Rollout around the VAB to the launchpad took place on March 11, 1969. (J.L. Pickering.)

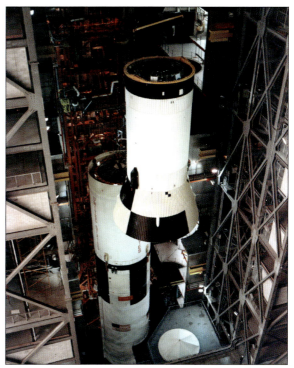

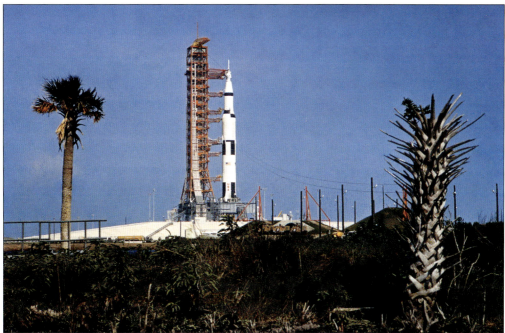

Three weeks before the launch of Apollo 10, a largely unreported incident happened with the rocket on the launchpad. On April 27, 1969, the fuel pre-valves on the S-IC-5 first stage were opened inadvertently, creating a negative pressure that caused the upper bulkhead of the fuel tank to collapse. The dome was popped back by applying pressure, but this left wrinkles in three of the gore segments. This photograph was taken on May 17, 1969. (AS10-S69-34328. J.L. Pickering.)

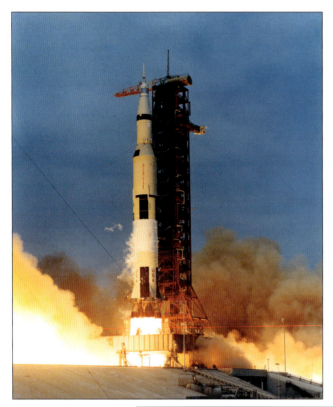

Apollo 10 became the only Saturn V to be launched from Launchpad 39B at KSC. It left the pad at 12:49 p.m. Eastern Daylight Time on May 18, 1969. No major anomalies occurred during the Apollo 10 launch. While the first and second stages fell into the ocean, the third stage is still in free flight with an indefinite lifetime in a heliocentric orbit. (AS10-69PC-168. J.L. Pickering.)

This fish-eye lens photograph shows the launch of Apollo 10 with astronauts Tom Stafford, Gene Cernan, and John Young on board. They conducted a full dress rehearsal for the first moon landing, with the LM coming within 47,400 feet of the lunar surface. (J.L. Pickering.)

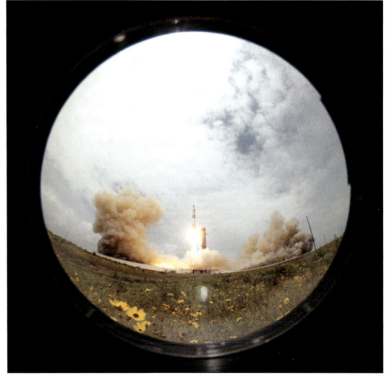

Six

MOON LANDINGS AND BEYOND

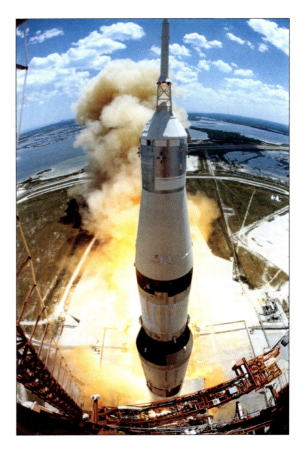

Saturn Vs launched six successful moon landing missions that resulted in 12 men walking on the moon between 1969 and 1972. Pres. John F Kennedy's goal of landing a man on the moon by the end of the decade was achieved. In addition, the Apollo 13 launch was successful, but the mission later was aborted, and the final Saturn V launched the Skylab space station. This image shows the launch of Apollo 16 on April 16, 1972. (AS16-72PC-172. J.L. Pickering.)

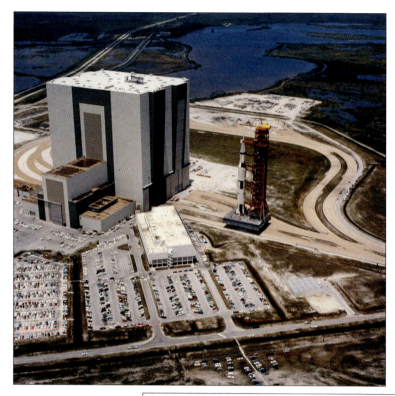

The Apollo 11 SA-506 launch vehicle was transferred from the VAB to Launchpad 39A on May 20, 1969. Apollo 11 was the final Saturn V to be stacked on ML-1. Two days later, the MSS was moved up to the rocket to provide technicians access to all parts of the rocket during launch preparations. (AS11-69HC-619. J.L. Pickering.)

Wernher von Braun was the head of the Marshall Space Flight Center in Huntsville throughout the period of the development and production of the Saturn V. Although the Saturn V was produced and tested by industry, it was the government, in the form of NASA, that managed the work. Dr. von Braun is seen here with the Apollo 11 rocket on Launchpad 39A at KSC in the lead-up to the launch. (69-01046. Kipp Teague.)

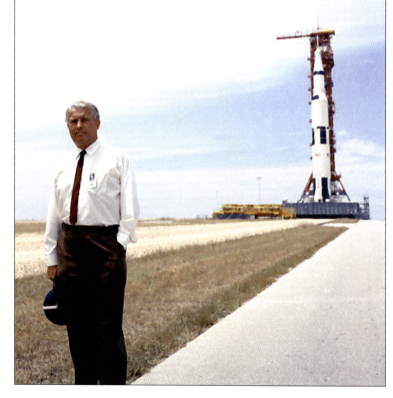

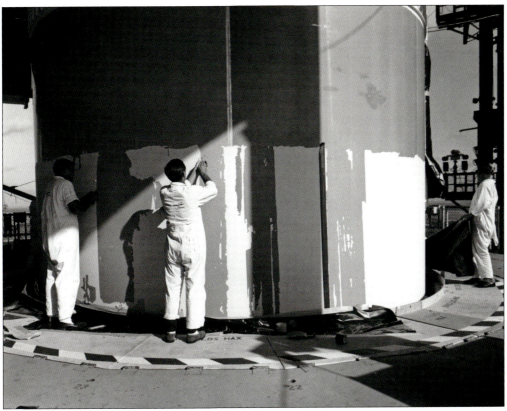

On July 5, 1969, just under two weeks from launch, a McDonnell Douglas team was sent up to the work platform around the Apollo 11 S-IVB third stage to perform an unusual task. The white paint on the outside of the third stage had started to peel, and it was deemed necessary to remove and repaint the stage at this very late stage in the launch preparation. (AS11-0708-69P-564. J.L. Pickering.)

Five days before launch, the Apollo 11 Saturn V sat on Launchpad 39A in the glow of the arc lights. For this flight and onwards, the R&D instrumentation was removed, saving weight and increasing payload capacity. The extra instrumentation was needed on the earlier flights in order to record all the various performance parameters of each stage. (AS11-69PC-357. J.L. Pickering.)

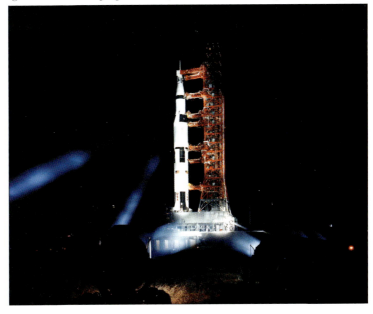

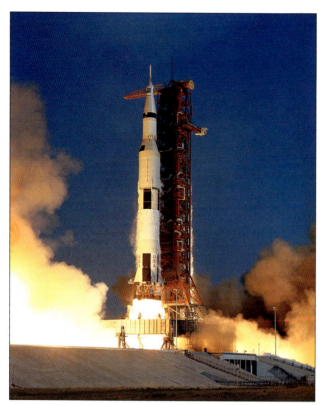

Launch of Apollo 11 occurred at 9:32 a.m. Eastern Daylight Time on July 16, 1969, transporting astronauts Neil Armstrong, Buzz Aldrin, and Mike Collins to the moon for the first lunar landing. NASA utilized 201 automatic cameras to record the launch from close quarters. Of these, 119 were for engineering purposes and 82 for documentary coverage. Only three cameras failed to work properly. (J.L. Pickering.)

Cameras were located within protected boxes at various positions on the LUT in order to image the vehicle during the first seconds of launch. The engine start sequence command was sent 8.9 seconds before lift-off, with the five engines being up to full thrust 1.6 seconds before launch. (J.L. Pickering.)

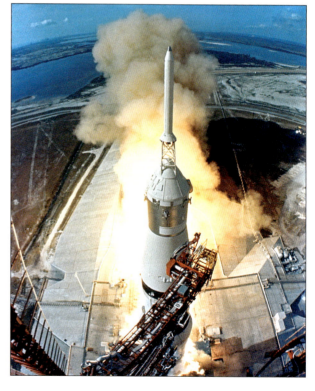

This is another view from the LUT and shows the first stage and F-1 engines passing the camera. The rocket reached Mach 1, the speed of sound, after 66.3 seconds at an altitude of 4.8 miles. (AS11-KSC-69PC-419. J.L. Pickering.)

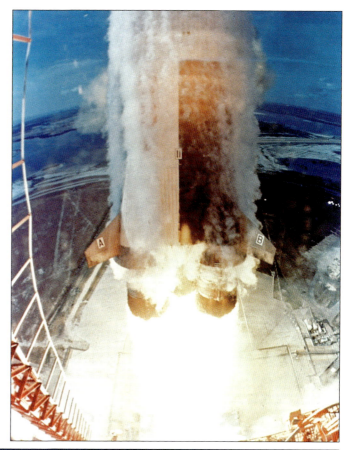

Separation of the first and second stages occurred 162 seconds after launch at a speed of 5,375 miles per hour. This tracking camera view clearly shows bright flashes near the base of the S-IC first stage where the four pairs of retro-rockets had just fired. Each solid-propellant retro-rocket had a thrust of 88,600 pounds and burned through the fairing that covered each F-1 engine. (AS11-S69-39958. J.L. Pickering.)

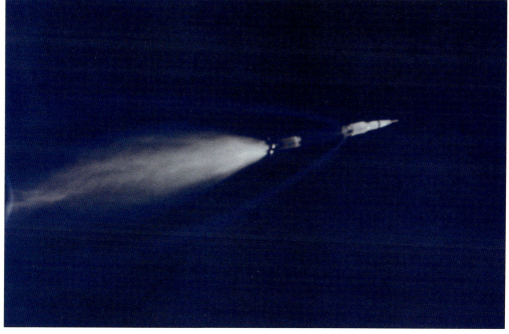

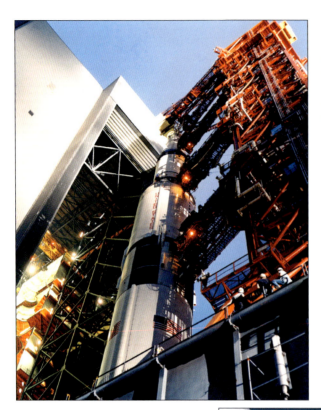

The 363-foot-tall Apollo 12 Saturn V, SA-507, exits HB3 of the VAB as the stack on top of ML-2 heads out to Launchpad 39A on September 8, 1969. The Crawler Transporter carried the 12.8-million-pound load 3.5 miles to the launchpad at a speed of less than one mile per hour. (AS12-69PC-529. J.L. Pickering.)

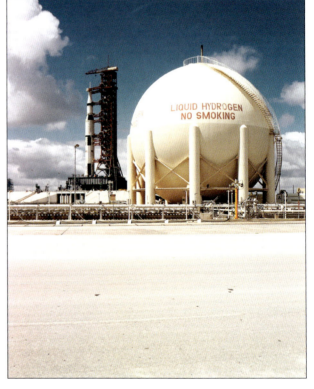

At Launchpad 39A, there were dedicated single fuel and oxidizer propellant storage tanks, each located 1,450 feet from the pad. This was determined as being a safe distance in the event of an explosion on the pad. In addition, there were three smaller RP-1 kerosene fuel tanks that were required for the first stage only. The liquid hydrogen tank, seen in this Apollo 12 image, had a capacity of 850,000 gallons. (AS12-69PC-532. J.L. Pickering.)

The Apollo 12 launch took place at 11:22 a.m. Eastern Standard Time on November 14, 1969. It carried the crew of Pete Conrad, Al Bean, and Dick Gordon on the second moon landing. The rocket passed through a charged electric field created by a cold front. The Saturn V rocket was struck by lightning at 36.5 seconds after the launch and again at 52 seconds. (AS12-S69-59475. J.L. Pickering.)

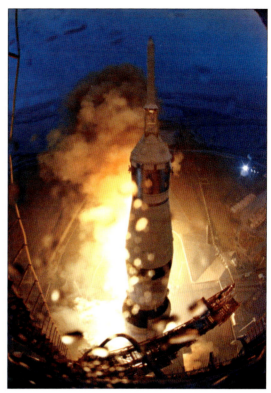

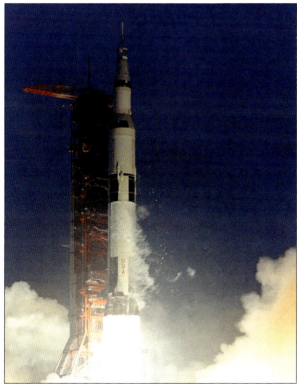

The length of the rocket and its exhaust plume triggered the lightning, which, in the case of the initial strike, also hit the launch tower. A call from John Aaron in Mission Control to switch the Signal Conditioning Equipment (SCE) to a backup power supply, "SCE to aux," together with Al Bean's recognition of this command, restored the spacecraft's three fuel cells, which had been disabled by the strike. (AS12-69PC-654. J.L. Pickering.)

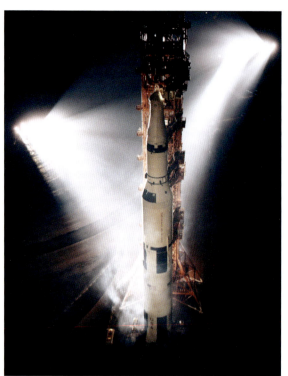

The Apollo 13 Saturn V, SA-508, was the second vehicle to be stacked on ML-3. The assembly was transferred to Launchpad 39A on December 15, 1969, pictured here rolling out of the VAB illuminated by predawn floodlights. (AS13-70PC-176. J.L. Pickering.)

Apollo 13, complete with a Boilerplate Command Module (CM), had originally been stacked in HB2, on the west side of the VAB, but was moved around the outside of the VAB to HB1, facing the launchpad, on August 8, 1969, after which the Boilerplate was replaced with the real CM. This was the final time that the HB1 door was opened to send a Saturn V to the launchpad. (AS13-69PC-820. J.L. Pickering.)

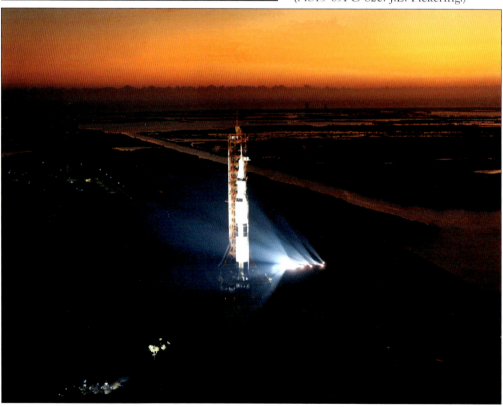

Apollo 13 was launched at 2:13 p.m. Eastern Standard Time on April 11, 1970, with a crew of Jim Lovell, Fred Haise, and Jack Swigert. Because of high-amplitude, low-frequency oscillations, the S-II center engine, number five, cut off 132 seconds earlier than planned. The remaining four engines burned for 35 seconds longer than planned to compensate. The problem was ultimately caused by a reduction in the LOX ullage pressure of three pounds per square inch. (AS13-70HC-355. J.L. Pickering.)

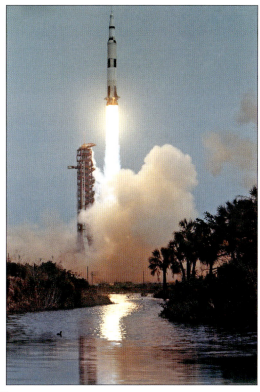

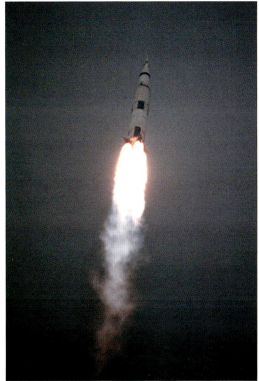

The Apollo 13 lunar landing was aborted due to an explosion in a Service Module (SM) oxygen tank. The S-IVB third stage became the first Saturn stage to be purposely crashed onto the lunar surface. It impacted the moon on April 14, 1970, and the seismic shock waves were picked up by the instruments left by the previous two crews. (J.L. Pickering.)

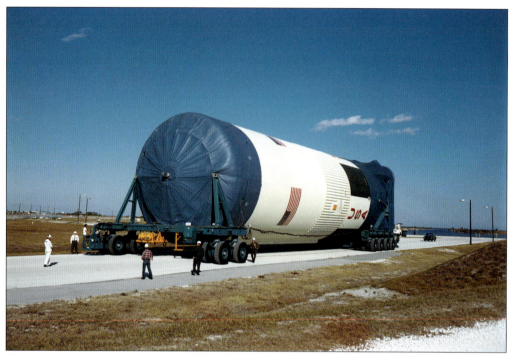

The first stage for the Apollo 14 SA-509 vehicle, designated S-IC-9, was shipped from Michoud on January 6, 1970, following the usual static testing. The covered barge arrived at the KSC turning basin and offloaded the stage on January 12, 1970, as seen in this image. (AS14-S70-25026. J.L. Pickering.)

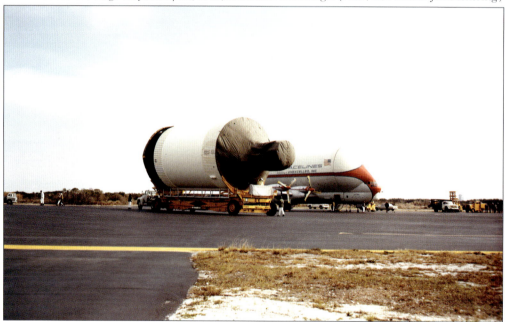

Following the static firing of the Apollo 14 S-IVB-509 third stage at SACTO, the stage was flown by the Super Guppy cargo aircraft to KSC. The aircraft took off from Mather Air Force Base near Sacramento on January 17, 1970, and, after several refueling stops, touched down at the KSC skid strip on January 20, 1970. (AS14-70PC-5. J.L. Pickering.)

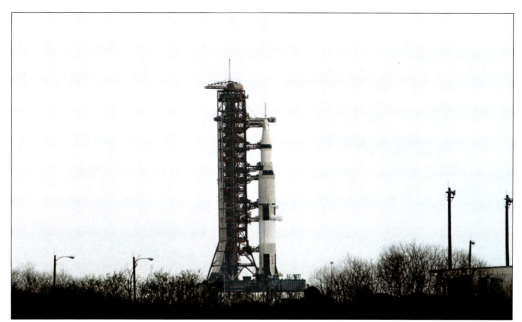

Apollo 14 was rolled out to Launchpad 39A on November 9, 1970, and launched at 4:03 p.m. Eastern Standard Time on January 31, 1971, with a crew of Alan Shepard, Ed Mitchell, and Stuart Roosa. It was the final manned Saturn V to be stacked on ML-2. The crew performed the third lunar landing, which was a replication of the planned mission aborted by Apollo 13. (J.L. Pickering.)

This view from an automatic camera on the LUT shows a close-up of the S-II second stage passing the camera during launch. At T-minus 8 minutes and 2 seconds, there was a 40-minute, 2-second hold due to high overcast clouds and rain. Following the Apollo 12 lightning strike, additional weather constraints had been imposed for safety reasons. Following the Apollo 13 S-II engine problem, a LOX feedline accumulator was added as a pogo suppressor. (J.L. Pickering.)

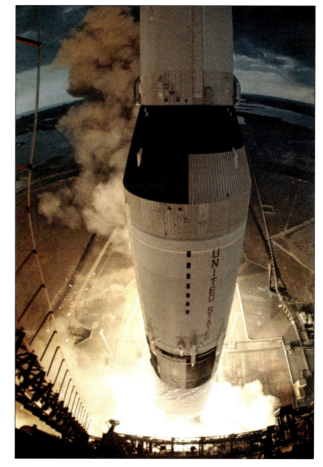

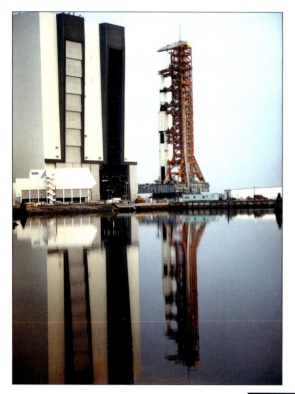

The three final manned Saturn V rockets were all assembled on ML-3 in HB3 in the VAB. The Apollo 15 Saturn V, SA-510, was rolled out of the VAB towards Launchpad 39A on May 11, 1971. Since Apollo 11, lower than predicted thrust had been obtained from the first stages during ground test. Consequently, on Apollo 15, as it was carrying a Lunar Rover, the engines were re-orificed for higher thrust. (AS15-71-33786. J.L. Pickering.)

The first stage engine re-orificing was done after the stage static firing, which was unique, unusual, and could be said to be risky. One engine was replaced completely after the stage test. Other changes included deletion of four of the eight S-IC retro-rockets and deletion of all the S-II ullage rockets. Again, these changes were made to reduce weight because of the heavier Apollo spacecraft. (AS15-71PC-568. J.L. Pickering.)

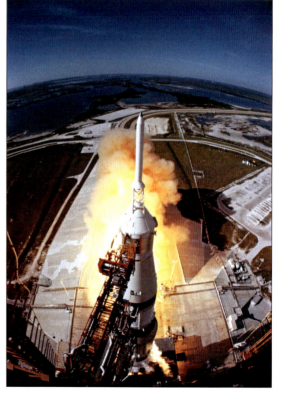

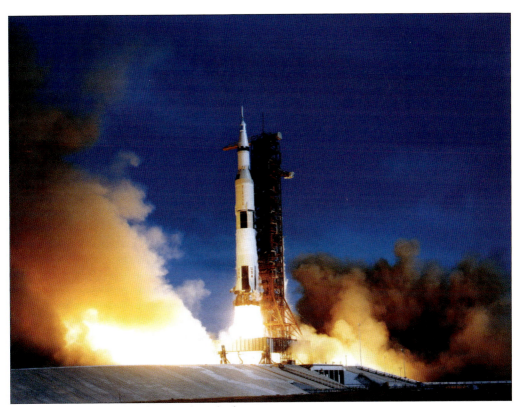

The combination of the changes described in the two previous captions together with an unexpected slow tail-off of thrust at first stage engine shut-off resulted in a reduced separation from the first and second stages, with a collision between the two stages almost occurring. Subsequent missions reverted to having eight first-stage retro-rockets. (AS15-71PC-571. J.L. Pickering.)

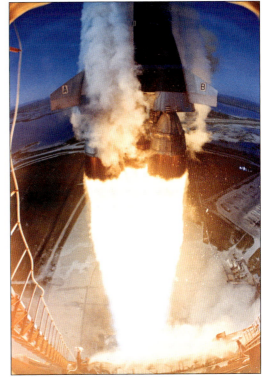

Apollo 15 was launched at 9:34 a.m. Eastern Daylight Time on July 26, 1971, with a crew of Dave Scott, Jim Irwin, and Al Worden. It was the first J-type mission, with an advanced lunar payload, and included the first use of a Lunar Rover. This image and the two previous ones show the Apollo 15 launch from different viewpoints. (AS15-71PC-560. J.L. Pickering.)

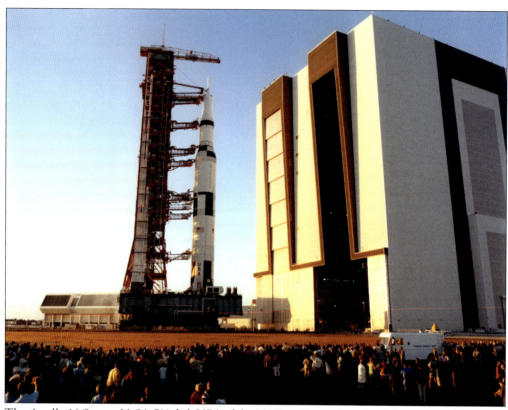

The Apollo 16 Saturn V, SA-511, left HB3 of the VAB on December 13, 1971, supported by ML-3. The first stage had undergone a yearlong refurbishment and engine replacement following the fire during the first acceptance test firing at MTF in June 1969. (AS16-71PC-768. J.L. Pickering.)

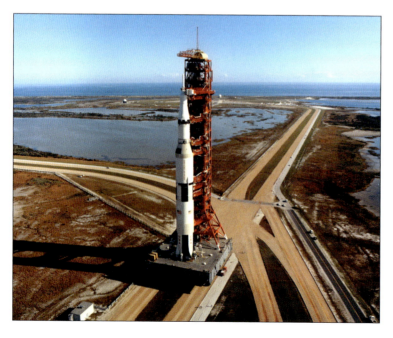

This is a view of the rollout from the roof of the VAB on December 13, 1971. For the first time since SA-500F, and unique to the Saturn V manned program, this launch vehicle had to be transported back to the VAB for some repair work. A Teflon diaphragm in a CM propellant tank burst, meaning that the tank had to be replaced, and access could not be obtained at the pad. (AS16-71PC-766. J.L. Pickering.)

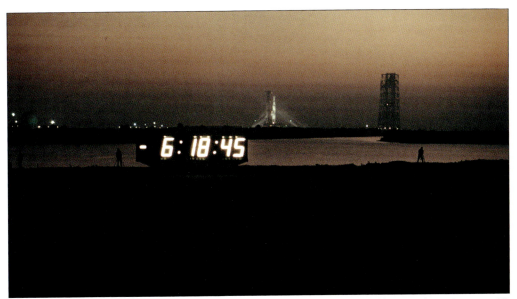

The stack was returned to the VAB for replacement of the CM propellant tank on January 27, 1972, and returned to the pad on February 9, 1972. This resulted in the launch date slipping from March 17 to April 16, 1972. This view from the press site shows the famous countdown clock on the morning of the launch. (Usciak Archives/J.L. Pickering.)

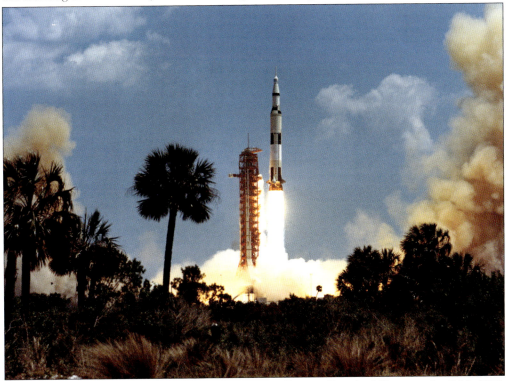

The Apollo 16 Saturn V was launched at 12:54 p.m. Eastern Standard Time on April 16, 1972, with astronauts John Young, Charlie Duke, and Tom Mattingly. The mission was the second J-type mission and carried the second Lunar Rover on board the LM. (AS16-S72-35345. J.L. Pickering.)

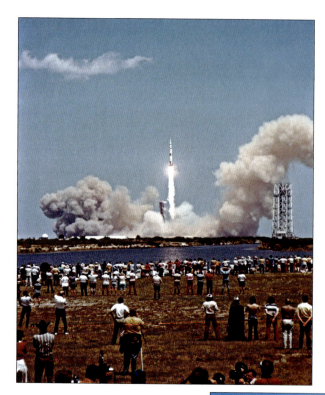

Here is another view of the Apollo 16 launch as seen from the press site. Five hours and 40 minutes into the mission, following the two firings of the S-IVB J-2 engine, there was a 54-second burn of the S-IVB auxiliary propulsion system to target the S-IVB for impact with the moon. However, a transponder failure resulted in lost telemetry that prevented determination of the precise impact point (see page 94). (J.L. Pickering.)

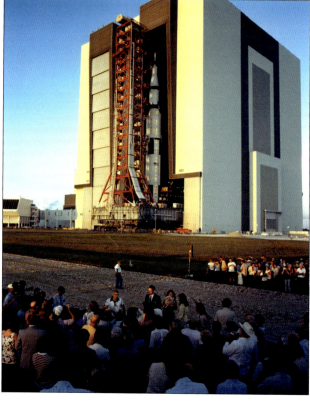

The Apollo 17 Saturn V, SA-512, was the final manned Saturn V launch vehicle. Erection of the launch vehicle on ML-3 had started on May 15, 1972. The rocket was rolled out from HB3 of the VAB in the early morning of August 28, 1972. The commander, Gene Cernan, can be seen in the foreground answering media questions. (AS17-72PC-441. J.L. Pickering.)

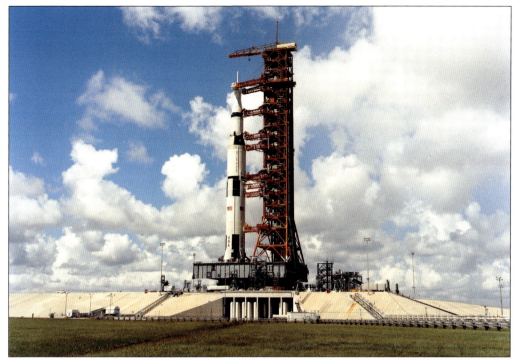

This view shows the Apollo 17 rocket after arrival at Launchpad 39A. The unique aspect of this launch vehicle was that the S-IVB third stage was never test fired on the ground. Due to budgetary constraints, the expensive static firings of this and later (not flown) S-IVB stages were deleted, adding somewhat to the risks during launch. (AS17-72PC-431. J.L. Pickering.)

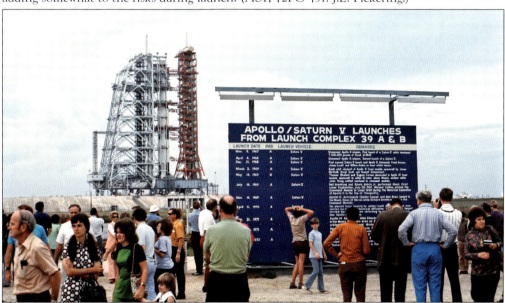

On December 5, 1972, just over a day before launch, tour busses were still taking visitors to view the launch record board at Launchpad 39A, a short distance from the Saturn V, which at that time was still surrounded by the MSS. The MSS later would be rolled back to a holding point along the crawlerway that leads back to the VAB. (J.L. Pickering.)

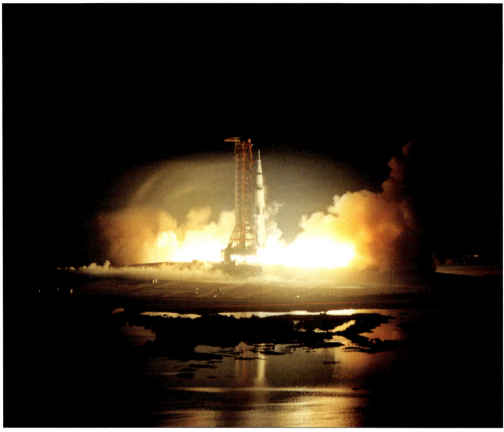

At T-minus 167 seconds, the Terminal Countdown Sequencer (TCS) failed to issue a command which, although manually overridden, resulted in an interlock relay not being energized. This led to a hold in the countdown at T-minus 30 seconds that ultimately lasted two hours and 40 minutes. This delay included another hold at T-minus 8 minutes after the countdown had been recycled to T-minus 22 minutes. (AS17-72PC-640. J.L. Pickering.)

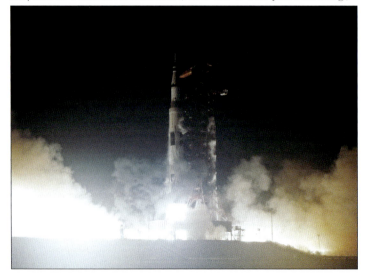

Apollo 17 was launched at 12:33 a.m. Eastern Standard Time on December 7, 1972, taking astronauts Gene Cernan, Jack Schmitt, and Ron Evans to the sixth and final moon landing. The launch was unique for the Saturn V as it was the only night-time launch and was visible to ground observers far up the East Coast of the United States. (AS17-72PC-599. J.L. Pickering.)

This photograph shows the final manned launch of a Saturn V rocket on December 7, 1972, as viewed by an automatic camera near the top of the LUT of the Mobile Launcher. When the author asked Gene Cernan about the lack of testing on this S-IVB stage, he replied that the crew had not been informed. (Kipp Teague.)

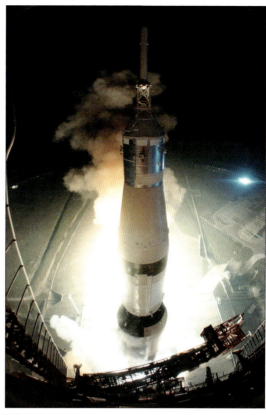

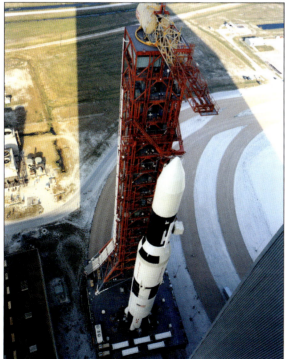

SA-513 was the 13th and final Saturn V to be launched. All launches were successful. For this mission, the third stage and Apollo payload were substituted by the Skylab space station, the main part of which, the Orbital Work Shop (OWS), was a converted Saturn IB S-IVB second stage built six years previously. Rollout on ML-2 from HB2, around the back of the VAB, took place on April 16, 1973. (SL1-73PC-184. J.L. Pickering.)

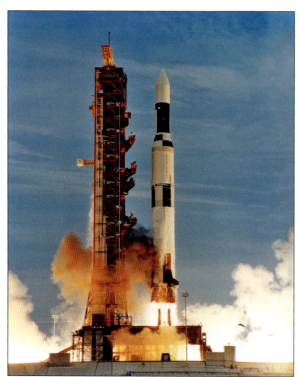

The Skylab 1 SA-513 rocket was launched from Launchpad 39A at 1:30 p.m. Eastern Daylight Time on May 14, 1973. It was the third unmanned flight of the Saturn V rocket, after Apollos 4 and 6. As the payload, the Skylab space station, had no propulsion capability itself, the S-II second stage was needed to take Skylab all the way up to orbital altitude and consequently did not decay from orbit for 606 days. (SL1-73PC-284. J.L. Pickering.)

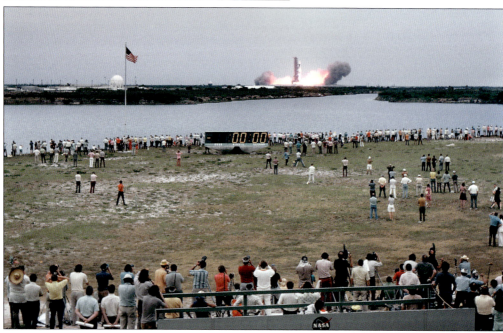

The countdown clock reaches zero as the final Saturn V to be launched leaves the ground. Sixty-three seconds into the flight, structural failure caused the micrometeoroid shield to come away. Debris released resulted in the S-IC/S-II inter-stage ring failing to separate and caused the OWS solar array no. 2 to partially deploy. When the S-II retro-rockets fired after 593 seconds, the exhaust plume tore the damaged wing completely off. (Usciak Archives/J.L. Pickering.)

Seven

AFTERMATH

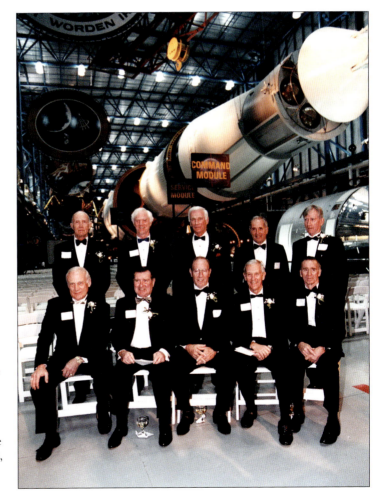

After the Apollo and Skylab missions were completed, a number of Saturn V flight and test stages remained. These were allocated to museums, and over the years, most have been restored. The first collection of stages to be restored were those at KSC, which were celebrated in the new Apollo/Saturn V Center during a gala evening on January 8, 1997, attended by, from left to right, (first row) astronauts Buzz Aldrin, Richard Gordon, Edgar Mitchell, Charles Duke, and Walter Cunningham; (second row) Thomas Stafford, Rusty Schweickart, Gene Cernan, William Anders, and John Young. (KSC-97PC-0130. NASA.)

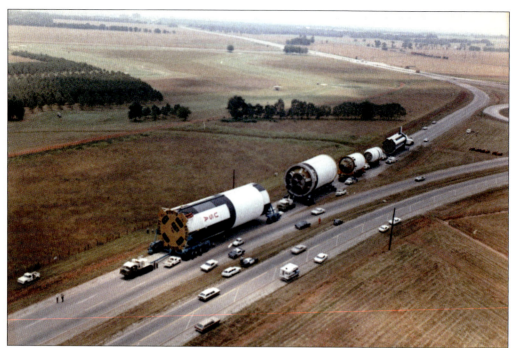

The Alabama Space and Rocket Center, later called the USSRC, was opened in Huntsville in 1970. Forthcoming Saturn rocket exhibits were moved from the nearby MSFC on June 28, 1969, as shown in this image. From left to right are the S-IC-D Dynamics stage, the S-II-F/D Facilities and Dynamics stage, the S-IVB-D Dynamics stage, the S-IV stage, and the S-I/IB stage. (69-01270. Arlene Royer [NARA].)

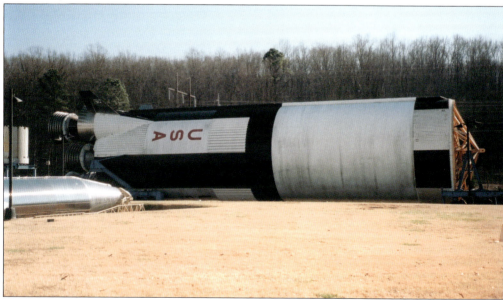

The S-IC-D stage was displayed outdoors at the USSRC for over 35 years. It was never designed to be test fired and had dummies installed in the engine locations. Over the years, the weather took its toll, although the stage was still an impressive sight at the time of this January 2004 photograph. (Alan Lawrie photograph.)

Between 2005 and 2007, a complete restoration of the three Saturn V stages at the USSRC took place. This photograph, taken in August 2005, shows the S-II-F/D stage during the strip down, repair, and repainting process. (USSRC.)

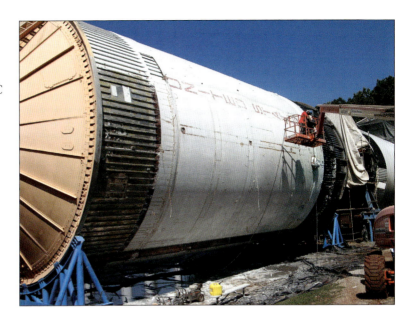

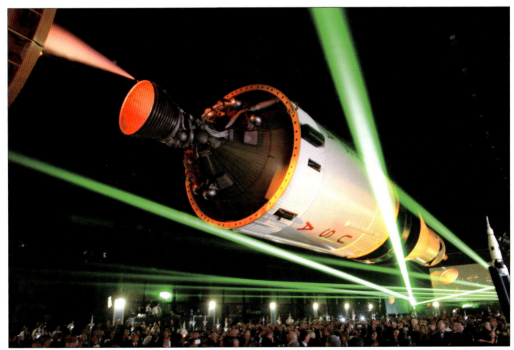

As well as the restoration of the three Saturn V stages in Huntsville, a magnificent new building, the Davidson Center, was constructed to house the rocket stages. Each stage is elevated, allowing wonderful viewing opportunities. The official opening of the Davidson Center, with a VIP dinner under the rocket, took place on January 31, 2008, which was the 50th anniversary of the launch of the first US satellite, Explorer 1. (USSRC.)

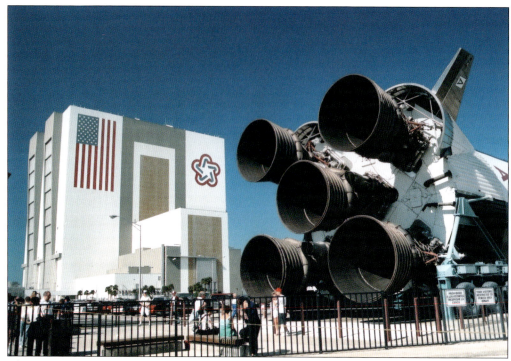

Three Saturn V stages were placed on display, in the open, next to the VAB at KSC in Florida at the time of the Bicentennial celebrations in 1976. This March 1991 photograph shows the F-1 engines at the rear of the first stage in front of the VAB. This stage is the S-IC-T, which was the original first stage to be test fired at MSFC in 1965. (Alan Lawrie photograph.)

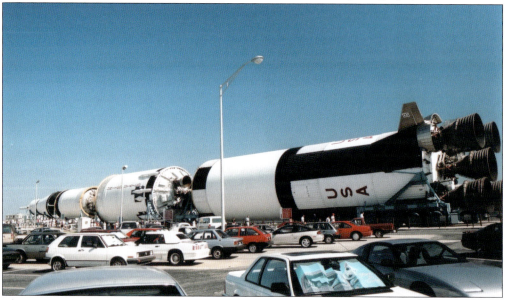

This is another March 1991 view of the KSC Saturn V when displayed outdoors. It shows the complete rocket, which in addition to the S-IC-T test stage included the S-II-14, an unused flight version second stage, and the S-IVB-514, an unused flight version third stage. (Alan Lawrie photograph.)

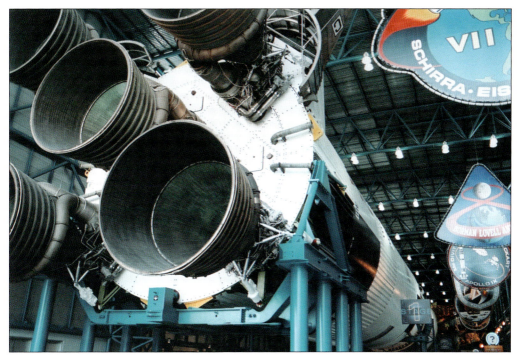

Twenty years in the hot Florida sun and salty atmosphere took its toll, and the KSC Saturn rockets were in need of restoration. The three stages were fully restored and put on display in a new Apollo/Saturn V Center that opened on December 5, 1996. The first stage was painted to mimic the Apollo 11 S-IC-6 stage, although it is in fact the S-IC-T stage. This photograph is from January 2001. (Alan Lawrie photograph.)

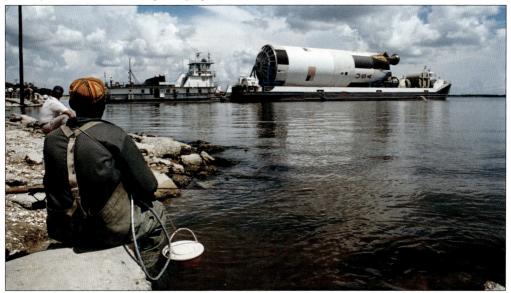

The NASA Johnson Space Center has its own display of Saturn V rocket stages. This collection includes the only S-IC first stage intended for flight that is on display for the public. This photograph, taken on September 19, 1977, shows the S-IC-14 stage arriving at JSC by barge following a journey from Michoud via the Intracoastal Waterway. (S77-28447. J.L. Pickering.)

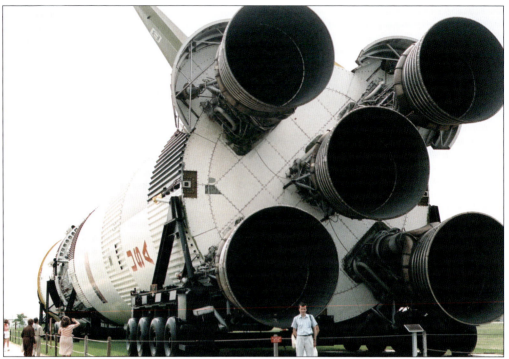

The Saturn V stages were put on display for the public in the open at the JSC visitor center. This March 1991 photograph shows the author next to the S-IC-14 first stage. Ahead of this are the S-II-15 second stage and the S-IVB-513 third stage. It is the only display anywhere that includes three Saturn V stages originally intended for flight. (Olwyn Lawrie.)

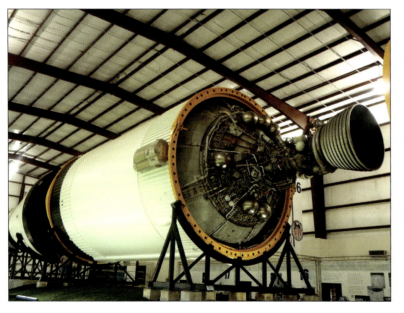

The JSC Saturn V also underwent restoration, which began in 2004 and was completed on July 20, 2007, when the refurbished rocket was unveiled in a purpose-built structure in the JSC visitor area. This May 2015 photograph shows the J-2 engine at the end of the S-IVB-513 stage. (Andy Clark.)

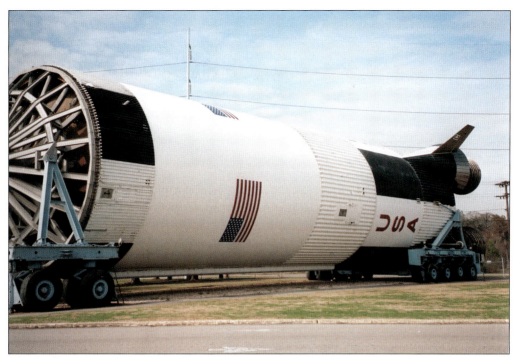

The final Saturn V first stage to be constructed, S-IC-15, was located inside the grounds of the Michoud Assembly Building, where it was built in 1970. Its only operational use was to act as back-up first stage for the Skylab-1 launch vehicle, but it was never needed. Although it was not possible for the public to view the stage close-up, it could be seen through the perimeter fence. This January 2004 image shows the stage where it had been from 1978 until June 2016. (Alan Lawrie photograph.)

This January 2004 view shows the F-1 engines attached to the S-IC-15 stage at Michoud. The stage includes the last F-1 engine ever made and also, interestingly, includes two engines that had previously been part of the S-IC-11 stage during its aborted firing in June 1969 (see page 20). As this is the only Saturn V stage that has not undergone restoration, there is currently an initiative to fund restoration. (Alan Lawrie photograph.)

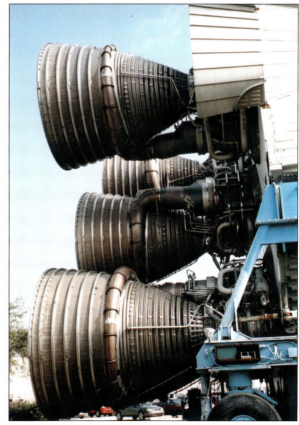

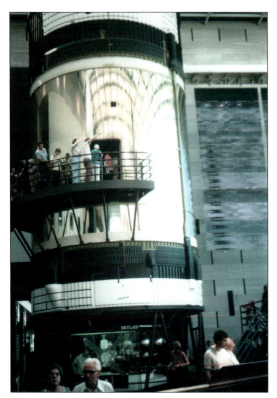

A back-up Skylab OWS was constructed from the final S-IVB stage to be built, S-IVB-515. After the completion of the Skylab mission, the OWS was shipped by barge from KSC to Washington, DC, in three separate pieces. It was reassembled in the National Air and Space Museum and has been on display ever since the museum opened in July 1976. This photograph was taken in April 1977. (Alan Lawrie photograph.)

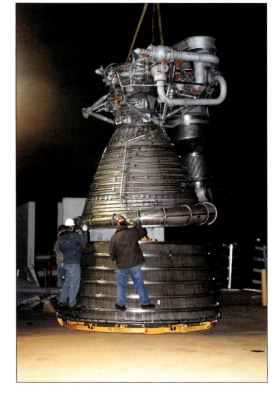

A number of individual F-1 and J-2 engines can be seen in museums around the world. Some of these were built for flight and were spares, and others were development engines. Other Saturn engines are at government establishments such as MSFC and have been examined in the frame of the Space Launch System (SLS) project. This image shows the assembly of the two halves of the F-5036 F-1 engine at the MSFC Administration Building display area in May 2008. (Doug Galloway.)

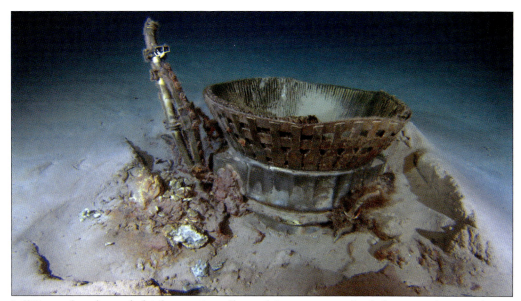

Jeff Bezos, founder and chief executive officer of Amazon.com, funded an expedition to locate and recover F-1 engines from the flown Saturn V rockets, which have been at the bottom of the Atlantic Ocean since the Apollo launches took place. David Concannon led the expedition, which in March 2013 photographed and then recovered parts from various F-1 engines. An example is this F-1 thrust chamber on the ocean floor. (Bezos Expeditions.)

After more than two years of careful conservation, the first example from the haul of recovered F-1 engine parts was put on public display. An Apollo 12 F-1 engine injector was uncovered by Allison Loveland of Seattle's Museum of Flight on November 19, 2015. Geoff Nunn, adjunct curator for space history (left), and Jeff Bezos (right) look on. (Dieter Zube.)

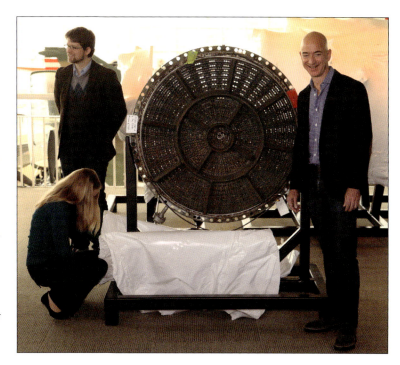

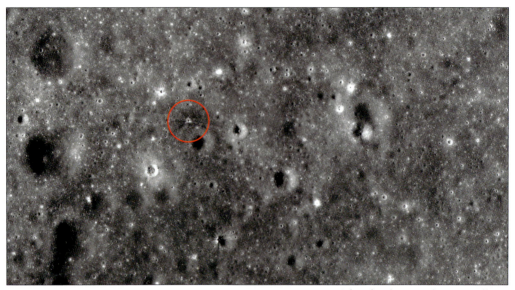

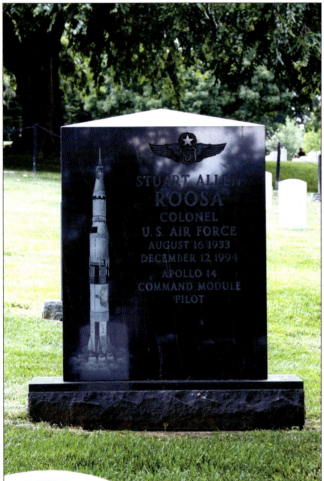

The third stages from Apollos 13 to 17 were all purposely crashed onto the lunar surface. The craters that they produced were photographed and identified by NASA's Lunar Reconnaissance Orbiter, except for that from Apollo 16 because of lost telemetry (see page 80). However, in November 2015, it was announced that the Apollo 16 crash site had been found. It produced a crater about 130 by 100 feet with a large mound in the middle. (M183689432L. NASA/Goddard/Arizona State University.)

In total, 26 men were launched by Saturn V rockets. One of those, Apollo 14 Command Module pilot Stuart Roosa, passed away in 1994. His gravestone, seen in this June 2007 photograph, has a wonderful engraving of a Saturn V rocket. (Mike Jetzer.)

BIBLIOGRAPHY

Analysis of Apollo 12 Lightning Incident. NASA TM-X-62894. February 1970.

Lawrie, Alan. *Sacramento's Moon Rockets*. Charleston, SC: Arcadia Publishing, 2015.

Lawrie, Alan, with Robert Godwin. *Saturn V: The Complete Manufacturing and Test Records*. Burlington, ON: Apogee Books, 2005.

Saturn V Launch Vehicle Flight Evaluation Report: AS-501. Apollo 4 mission. NASA MPR-SAT-FE-68-1. January 1968.
Saturn V Launch Vehicle Flight Evaluation Report: AS-502. Apollo 6 mission. NASA MPR-SAT-FE-68-3. June 1968.
Saturn V Launch Vehicle Flight Evaluation Report: AS-503. Apollo 8 mission. NASA MPR-SAT-FE-69-1. February 1969.
Saturn V Launch Vehicle Flight Evaluation Report: AS-504. Apollo 9 mission. NASA MPR-SAT-FE-69-4. May 1969.
Saturn V Launch Vehicle Flight Evaluation Report: AS-505. Apollo 10 mission. NASA MPR-SAT-FE-69-7. July 1969.
Saturn V Launch Vehicle Flight Evaluation Report: AS-506. Apollo 11 mission. NASA MPR-SAT-FE-69-9. September 1969.
Saturn V Launch Vehicle Flight Evaluation Report: AS-507. Apollo 12 mission. NASA MPR-SAT-FE-70-1. January 1970.
Saturn V Launch Vehicle Flight Evaluation Report: AS-508. Apollo 13 mission. NASA MPR-SAT-FE-70-2. June 1970.
Saturn V Launch Vehicle Flight Evaluation Report: AS-509. Apollo 14 mission. NASA MPR-SAT-FE-71-1. April 1971.
Saturn V Launch Vehicle Flight Evaluation Report: AS-510. Apollo 15 mission. NASA MPR-SAT-FE-71-2. October 1971.
Saturn V Launch Vehicle Flight Evaluation Report: AS-511. Apollo 16 mission. NASA MPR-SAT-FE-72-1. June 1972.
Saturn V Launch Vehicle Flight Evaluation Report: AS-512. Apollo 17 mission. NASA MPR-SAT-FE-73-1. February 1973.
Saturn V Launch Vehicle Flight Evaluation Report: SA-513. Skylab-1. NASA MPR-SAT-FE-73-4. August 1973.

Discover Thousands of Local History Books Featuring Millions of Vintage Images

Arcadia Publishing, the leading local history publisher in the United States, is committed to making history accessible and meaningful through publishing books that celebrate and preserve the heritage of America's people and places.

Find more books like this at
www.arcadiapublishing.com

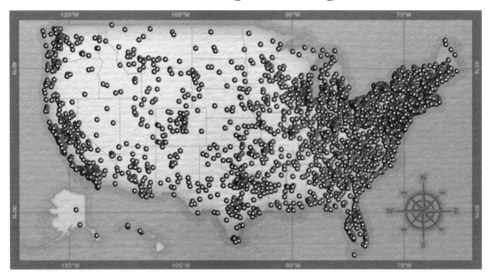

Search for your hometown history, your old stomping grounds, and even your favorite sports team.

Consistent with our mission to preserve history on a local level, this book was printed in South Carolina on American-made paper and manufactured entirely in the United States. Products carrying the accredited Forest Stewardship Council (FSC) label are printed on 100 percent FSC-certified paper.

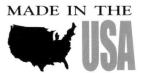